Science and Social Work

Science and Social Work

A CRITICAL APPRAISAL

Stuart A. Kirk and William J. Reid

COLUMBIA UNIVERSITY PRESS

New York

COLUMBIA UNIVERSITY PRESS
Publishers Since 1893
New York   Chichester, West Sussex

Library of Congress Cataloging-in-Publication Data
Kirk, Stuart A., 1945–
Science and social work : a critical appraisal /
Stuart A. Kirk and William J. Reid.
p.   cm.
Includes bibliographical references and index.
ISBN 0–231–11824–4 (cloth)
ISBN 0–231–11825–2 (pbk.)
1. Social service. 2. Science—Methodology.
3. Knowledge, Sociology of.
I. Reid, William James, 1928– II. Title.
HV40 .K462 2001
361.3—dc21   2001028014

∞

Columbia University Press books
are printed on permanent and durable acid-free paper.

Printed in the United States of America

c 10 9 8 7 6 5 4 3 2 1
p 10 9 8 7 6 5 4 3 2 1

This book is dedicated to two young experimentalists, bright and energetic, but not yet old enough to read:

LUCAS DYLAN KIRK and
MICHAEL STEVEN REID TWENTYMAN

■ CONTENTS

The title of this book may strike some people—and perhaps some social workers—as an oxymoron. The stereotypical white-coated scientist in a university basement laboratory, bent over a microscope and petri dish, might appear completely dissimilar from the standard image of a social worker finding her way through a public housing project to investigate a report about an abused child.

The scientist is physically separated from the social world, working in a specially constructed space designed to allow the study of natural phenomena under highly controlled conditions. He or she builds on the work of other scientists by following a highly formalized set of procedures to ensure that the inquiry is objective, verifiable, and unbiased. The knowledge being pursued may have very limited or no immediate applicability in the world outside the laboratory. The results of the experiment may appear in a scientific journal months later and be read by only a few other researchers. Yet the scientist's study may become a footnote in the work of successors, a small brick in the edifice perpetually under construction. Science is about many small discoveries, laboriously built on one another over the decades in the quest for knowledge.

The social worker operates in a world outside the university, contending with a flow of events, people, and problems that are far from controlled. In fact, social workers are often enlisted precisely because normal processes and structures have failed: children are not developing well; families are disintegrating; communities are disenfranchised. Social workers are called upon to help a particular client, to improve the situation, to make a difference today. They often must act without much time for contemplation, consultation, or research, intervening in ways that may appear unique to the particular circumstances or to the personal style and experience of the worker. The results of the intervention are rarely made public and are often difficult to assess objectively. Social work is about many small acts of kindness, persistently pursued in the quest to be of service to others.

From the earliest years of the profession, however, social work leaders have acknowledged the need to make use of science in two broad ways. One has been to use its rational, structured methods in delivering services to clients. Social work, they have suggested, could emulate science if practitioners gathered information more systematically, tried to formulate logical inferences, and carefully developed and monitored their interventions. That is, social workers could use the methods of science as techniques of practice. A second approach has been to use scientific evidence to inform social work activities. The products of science—the knowledge developed by researchers—could be used by social workers to better understand their clients' problems and to select inteventions most likely to be effective. That is, social work practice could be grounded in and guided by scientific knowledge.

Although the use of science in social work has rarely been a topic that has engaged or preoccupied the entire profession, it has been a persistent concern of small groups of prominent leaders, researchers, and scholars for a century. In different ways in different eras, dedicated advocates have called for closer collaboration between the worlds of science and social work practice. In this book, we trace the history and success of these varied efforts and offer a perspective on the enduring challenges that lie ahead.

In a sense, this is a personal as well as historical reflection. Both of us have spent much of our professorial lives thinking, teaching, and writing about the role of science and research in social work. It's certainly been an important, complicated, intriguing, and controversial endeavor. But it's been a frustrating one as well. We acknowledge that we are often preaching only to the choir at the university and in the academic journals, that we have made relatively few converts among practitioners, and that our efforts devoted to knowledge building seem small in relation to the enormity of the need for information and the tasks social workers undertake. Within the profession, science remains on the cultural margins, struggling for a voice and a following. In our critical appraisal of the efforts to link science and practice, we try to present a balanced and fair view, if for no other reason than that we have been active participants in many of these very efforts. Therefore, our criticisms of the work of others—many of whom are our friends and respected colleagues—are also, in part, self-criticism.

Since this book draws on our work over many decades, our gratitude for help received extends more widely than usual. Of course, we are grateful for the diffuse personal support that comes from our spouses, Carol Ann and Ricky, who intimately tolerate our peculiar habits of authorship. But we are also indebted to many people from many universities who have worked with us as coauthors, mentors, and colleagues over the years as we have struggled with the problems that we discuss in this book. Their names are scattered throughout the bibliography. By joining in our endeavors, many of our former students have kept the quest alive and satisfying. One, Carrie Petrucci, a Ph.D. candidate at UCLA, not only served as a coauthor of a chapter but also assisted us with unfailing industriousness on numerous other tasks. We are grateful, also, to our deans and department chairs, Dean Barbara Nelson and Professor Ted Benjamin at UCLA and Deans Lynn Videka-Sherman and Katharine Briar-Lawson at the State University of New York at Albany, for providing academic environments that respect and nurture scholarly inquiry.

In the end, however, our debt will be to the readers of this book, if they can understand what the scientist and the social worker have in common and if, in their own careers, they work to ensure that in the future, "science in social work" will no longer be viewed as a contradiction in terms.

Stuart A. Kirk
William J. Reid

The authors acknowledge, with thanks, permission to take and adapt some material from our prior publications: William J. Reid (1994), "The Empirical Practice Movement," *Social Service Review* 68:165–184; Stuart A. Kirk (1979), "Understanding Research Utilization in Social Work," in A. Rubin and A. Rosenblatt, eds., *Sourcebook on Research Utilization* (New York: CSWE), 3–15; Herb Kutchins and Stuart A. Kirk (1997), *Making Us Crazy: DSM: The Psychiatric Bible and the Creation of Mental Disorders* (New York: Free Press); William J. Reid and Anne Fortune (1992), "Research Utilization in Direct Social Work Practice," in A. Grasso and I. Epstein, eds., *Research Utilization in the Social Services* (Binghamton, NY: Haworth); Stuart A. Kirk (1990), "Research Utilization: The Substructure of Belief," in L. Videka-Sherman and W. J. Reid, eds., *Advances in Clinical Social Work Research* (New York: NASW Pres), 233–250; Sturat A. Kirk (1991), "Scholarship and the Professional School" in *Social Work Research and Abstracts* 27 (1): 3–5; Stuart A. Kirk, ed. (1999), *Social Work Research Methods* (New York: NASW Press); William J. Reid and P. Zettergren (1999), "Empirical Practice in Evaluation in Social Work Practice," in I. Shaw and J. Lishman, eds., *Evaluation in Social Work Practice* (Thousand Oaks, CA: Sage).

Science and Social Work

■ CHAPTER ONE

Knowledge, Science, and the Profession of Social Work

Just before noon on Monday, May 17, 1915, a young reformer from the General Education Board in New York City rose to speak at the 42nd annual meeting of the National Conference of Charities and Correction in Baltimore. The program for that day covered the topic "Education for Social Work." In his disarming opening remarks, the educator questioned his own competence to address the subject assigned to him, "Is Social Work a Profession?" because of his limited knowledge of social work and stressed that he was not prepared to press his points if they seemed unsound or academic. Although it is not known exactly how the audience that morning reacted, the speech caused a ringing in the ears of social workers that lasted nearly a century.

In the course of his speech, the educator, Abraham Flexner, concluded decisively that "social work is hardly eligible" for the status of a profession (Flexner 1915:588).[1] One reason for the notoriety of Flexner's claim was his reputation as a mover and shaker in professional education. He had just spearheaded an effort to reform medical education in the United States by

[1]Abraham Flexner was followed to the podium by a professor of law from Harvard University, Felix Frankfurter, who reinforced Flexner's conclusion by focusing on how the major professions had gravitated into universities, where there were both entrance and exit requirements and multiple-year graduate curricula. He argued that the tasks of an applied social science were likely to be no less demanding than those of law or medicine and urged the social work training schools to follow those professions' model of university affiliation (Frankfurter 1915).

purging many commercial, nonuniversity-affiliated schools. Medical education had been moving for decades from a diverse array of hundreds of proprietary schools with few if any admission or graduation requirements and only a brief curriculum to a smaller number of university-based programs, such as those at Harvard and Johns Hopkins. Even those two renowned institutions were asserting control over their autonomous medical schools and developing longer programs with more rigorous requirements. Johns Hopkins, for example, which had just opened its medical school in 1893, made two radical innovations: students had to have a college degree to enroll, and medical education would take an additional four years of study (Starr 1982). Furthermore, the Hopkins program, which provided the new blueprint for medical education nationally, joined science and research more firmly to clinical practice.

The American Medical Association was attempting to gain control over the numerous commercial medical schools through such mechanisms as state licensing boards and educational standards. In 1904, the AMA established a Council on Medical Education, which formulated minimum standards for medical education; when the Council began grading medical schools, it found many of them severely wanting in quality (Starr 1982:117–118). Fearing conflict among medical schools and professional organizations, the Council did not share its ratings outside of the medical fraternity. Instead, the AMA invited an outside group, the Carnegie Foundation for the Advancement of Teaching, to conduct an investigation of medical education. The Foundation in turn asked Flexner, whose brother was president of the Rockefeller Institute for Medical Research and who himself held a bachelor's degree from Hopkins, to lead the effort.

The results of his investigation, based on personal visits to each of the nation's medical schools, were released in 1910 and exposed many of the weaker proprietary schools to public embarrassment for grossly inadequate laboratories, libraries, faculty, and admission standards. In the words of medical historian Paul Starr, "As Flexner saw it, a great discrepancy had opened up between medical science and medical education. While science had progressed, education had lagged behind" (1982:120). The Flexner report hastened the decline in the number of medical schools and the number of medical doctors graduated each year, allowing the AMA to gain much greater control over education and practice. Flexner's report also directed the major foundations to invest heavily in a few leading research-oriented medical schools that had been assimilated into universities. As a consequence, medical education became dominated by scientists and researchers rather than by practitioners. This young reformer, not even a physician himself, had transformed medical education and practice in ways that are readily apparent to this day.

Flexner spoke before an audience of people interested in charities, settlement houses, corrections, and social work in 1915, and his reputation as a powerful figure in reforming professional education and professional practice was unquestioned. That is why his conclusion that social work was not yet a profession was given so much credibility. Although it has haunted the field for many decades, his reasons for reaching it are less well understood.

Flexner's Challenge

Flexner's conclusions about social work were not the results of any careful study, visits to social agencies, or interviews with faculty of the training schools; they were not based on a systematic review of the literature on professions or on any careful comparative methods. As his opening sentences indicate, he was aware that he knew relatively little about social work. His conclusions were derived from his consideration of what criteria ("certain objective standards") must be met for any occupation to rightfully claim professional status.

He begins by mentioning those occupations that were admitted to be professions—law, medicine, engineering, and preaching (and adds others, such as architecture and university teaching, in the course of his exposition). Then, by induction, he tries to identify the criteria for that designation. The first is:

> that the activities involved are essentially intellectual in character ... the real character of the activity is the thinking process. A free, resourceful, and unhampered intelligence applied to problems and seeking to understand and master them. (578)

Since professions are intellectual, the professional "thinker takes upon himself a risk" and a responsibility to exercise discretion (578). Mere routine instrumental or mechanical activity does not constitute a profession, because "some one back of the routineer has done the thinking and therefore bears the responsibility" (579).

The second criterion is that professions must be "learned," not largely employ knowledge "that is generally accessible" to everyone. Practitioners "need to resort to the laboratory and the seminar for a constantly fresh supply" of facts and ideas that keep professions from degenerating into mere routine and "losing their intellectual and responsible character" (579).

Third, professions must have some practical purpose. "No profession can be merely academic and theoretic"; rather, it must have "an absolutely definite and practical object." While professions may draw on the basic sciences, they "strive towards objects capable of clear, unambiguous, and concrete formulation" (579). Fourth, professions must be teachable in a curriculum, they must possess "a technique capable of communication through an orderly and highly specialized educational discipline" (580). Fifth, professions are "brotherhoods" of individuals who are selected based on their qualifications and who devote their lives to their work. Finally, Flexner argues that professions increasingly are organized for the achievement of social (not personal) ends, "the advancement of the common social interest" and the "devotion to well-doing" (581). He summarizes:

> Professions involve essentially intellectual operations with large individual responsibility; they derive their raw material from science and learning; this material they work up to a practical and definite end; they possess an educationally communicable technique; they tend to self-organization; they are becoming increasingly altruistic in motivation.   (581)

Flexner then uses these criteria to consider the professional status of various occupations, as a way to prove the validity of the criteria. For example, he concludes that plumbing is merely a handicraft because it is instrumental, not intellectually derived from science, and its purpose is profit, not social betterment. He faults banking as an insufficient application of "economic science" too focused on profit. He finds pharmacy and nursing wanting: both are not predominantly intellectual in character and the primary responsibility lies with the physician:

> It is the physician who observes, reflects, and decides. The trained nurse plays into his hands; carries out his orders; summons him like a sentinel in fresh emergencies; subordinates loyally her intelligence to his theory, to his policy, and is effective in precise proportion to her ability thus to second his efforts.   (583)

As if his argument is self-evidently valid, he rallies with this provocative statement:

> With medicine, law, engineering, literature, painting, music, we emerge from all clouds of doubt into the unmistakable professions. Without exception, these callings involve personally responsible intellectual activity; they derive their material immediately from learning and science; they possess an organized and educationally communicable technique; they have evolved into definite status, social and professional, and they tend to become, more and more clearly, organs for the achievement of large social ends. I need not establish this position separately in reference to each of them. (583)

Indeed, this is the only mention Flexner makes of literature, painting, or music. He picks the example most familiar to his recent reform efforts, medicine, and easily elaborates on how it meets his criteria.

He then turns his attention to social work, opening with a quote from the bulletin of the New York School of Philanthropy defining the field. He immediately concedes that social work appears to meet some of his criteria: the need for analysis, sound judgment, and skill "are assuredly of intellectual quality"; social workers derive their material from science and learning; they have a rapidly evolving professional self-consciousness; and they pursue broader social good, not personal profit.

Flexner fails social work on three other points. First, he suggests that social workers don't so much take final responsibility for the solution of a case (his requirement that "the thinker takes upon himself a risk") as perform a "mediating" function that brings other professions and institutions together to help but does not divide labor among equals—a conclusion similar to his criticism of pharmacy and nursing.

Second, he claims that social work lacks "definite and specific ends" and "appears not so much a definite field as an aspect of work in many fields." Its broad aims and diverse activities produce "a certain superficiality of attainment, a certain lack of practical ability," suggesting that it is "in touch with many professions rather than as a profession in and by itself" (586). Social work doesn't meet fully his third criterion of being directed toward an unambiguous objective.

Finally, he argues that with unspecific aims, social work has trouble providing a "compact, purposefully organized educational discipline" (587), thus failing to meet his fourth criterion. Social work requires resourceful, judicious, well-balanced people trained broadly in the "realms of civic and social interest" rather than in some "technically professional" program.

At the end, Flexner says that social work is "too self-confident," and along with journalism, suffers from excessive facility in speech and in action. Here he appears to be criticizing social workers for their zealous efforts at social reform. He then segues into a recommendation that social work not rely so heavily on newspapers for "news-propaganda and agitation," but consider developing a "dignified and critical means of expressing itself in the form of a periodical which shall describe in careful terms whatever work is in progress. . . . To some extent the evolution of social work towards the professional status can be measured by the quality of publications put forth in its name" (590).

His final comment emphasizes that the most important criterion is an unselfish devotion to making the world a better place. "In the long run, the first, main and indispensable criterion of a profession will be the possession of professional spirit, and that test social work may, if it will, fully satisfy" (590).

Although the customary summary of Flexner's article is that he faults social work for not having a body of scientific knowledge (see, for example, Thyer 2000), that is not precisely the case. In fact, he readily concludes that social work meets the criterion of having an intellectual quality based on science and learning. He is more troubled by the mediating role and broad, diffuse aims of social work, by its general efforts to do good and its limited autonomy, and by its primitive curriculum. An occupation's ambiguous aims might complicate the identification of the intellectual basis for action, but it is important to recognize how Flexner defines social work's primary problem.

The challenge that Flexner made to the developing profession—to clarify autonomous responsibility, refine aims, and develop training—might have been easy to address through legislation, licensing, and a more coherent professional purpose and curriculum, had it not been overshadowed by the charge that social work lacked a specific, separate scientific body of knowledge. It was this lack that became social work's accepted failing, the budding profession's Achilles heel.

Defining Professions

Flexner's criteria became one conventional and popular way of understanding the evolution of professions. A checklist of traits is derived inductively by comparing established professions, typically medicine, with other trades and inferring their distinguishing characteristics from apparent differences. That list of traits is then used to measure the extent to which various occupations over time progress in a linear fashion into full professional status.

But there are radically different approaches to understanding the evolution of professions, most notably an alternative theory offered by Andrew Abbott (1988, 1995), who suggests that professions must compete for jurisdiction in competitive and conflict-ridden arenas, not simply meet several objective criteria. Professions emerge as their members discover, create, mark, and maintain boundaries over occupational turf, sometimes left unattended or neglected by others and sometimes actively defended. For example, psychiatrists had to wrestle control of some human problems from the clergy and the law, and then later fend off encroachments by social workers, psychologists, and family counselors. Other budding professions—psychological medium and railroad dispatcher, for example—died when their tasks were assumed by others (Abbott 1988).

In contested territory, according to Abbott, abstract knowledge plays a somewhat different function.

The ability of a profession to sustain its jurisdiction lies partly in the power and prestige of its academic knowledge. This prestige reflects the public's mistaken belief that abstract professional knowledge is continuous with practical professional knowledge, and hence that prestigious abstract knowledge implies effective professional work. In fact, the true use of academic professional knowledge is less practical than symbolic. Academic knowledge legitimizes professional work by clarifying its foundations and tracing them to major cultural values. In most modern professions, these have been the values of rationality, logic, and science.

(1988:53–54)

Although Flexner and others assume that knowledge gained from the sciences and higher learning will be directly applied by skilled practitioners to clients' practical problems, thus giving a scientific aura to professional work, this assumption serves an important symbolic function by appearing to anchor professional activity in rationality rather than in tradition or legislative politics. This may be why the developers of the social work profession tended to focus more on the need to identify a "body of scientific knowledge" than on some of Flexner's other concerns.

Flexner's attempt to define professions via checklists has been duplicated many times over the years as the number of professions has grown and their influence in society has expanded. In "Attributes of a Profession" (1957)— perhaps the most-cited article about social work's professional status since Flexner—Ernest Greenwood, a Berkeley professor of social welfare, also outlines criteria for professional status. But unlike Flexner, he concludes that social work is a profession. By inductively reviewing those occupations that seem clearly to have achieved professional status, Greenwood enumerates and discusses what he considers to be the essential attributes of professions. Unlike Flexner, however, he does not then subject various striving occupations to his test; he doesn't even assess social work systematically by reviewing how and to what extent it meets each of his criteria. He merely concludes that it meets them all.

Greenwood's criteria overlap with Flexner's, although he uses different terms and develops each idea more systematically, as one would expect of a scholar who has had the benefit of sociological literature to draw on. Greenwood discusses the importance of systematic theory, professional authority derived from that specialized knowledge, the approval of the community to have a monopoly on practice, a regulatory code of ethics to govern relations with clients and colleagues, and a professional culture consisting of organizations, institutions, values, norms, and symbols. The first criterion—possession of a systematic body of theory—merits comment because he uses it to introduce the ongoing role of science and research in profession building.

Greenwood dismisses the argument that professionals are distinguished by superior skills and claims that the crucial distinction is whether practice skills "flow from and are supported by a fund of knowledge that has been organized into an internally consistent system, called a *body of theory*" (304; emphasis in original). "Theory serves as a base in terms of which the professional rationalizes his operations in concrete situations," so the student is required to master simultaneously both the theory and the skill derived from it (304). Preparation for a profession, then, entails "intellectual as well as a practical experience" (304). But the requirement of theory introduces an additional component:

> The importance of theory precipitates a form of activity normally not encountered in a nonprofessional occupation, *viz.*, theory construction via systematic research. To generate valid theory that will provide a solid base for professional techniques requires that application of the scientific method to the service-related problems of the profession. Continued employment of the scientific method is nurtured by and in turn reinforces the element of *rationality*. As an orientation, rationality is the antithesis of traditionalism. The spirit of rationality in a profession encourages a critical, as opposed to a reverential, attitude toward the theoretical system.   (305, emphasis in original)

We need to recall that at the time (the 1950s), social work was largely conviction, tradition, and untested practice theory. Greenwood was making a case for requiring that social work submit its "professional techniques" to the ongoing scrutiny of research and rationality. He argues that research is fundamental to theory and both are necessary for professional status.

One of the consequences of the need for systematic research, Greenwood suggests, is a division of labor "between the theory-oriented and the practice-oriented person" (305), for example, between the medical researcher and the private medical practitioner. At this time, social work had a very small cohort of theoreticians/researchers who were sheltered in a few elite universities, and although Greenwood concludes that social work had already gained the status of a profession, the resources dedicated to developing social work's body of theory were meager indeed compared to the rapidly developing medical research establishment. Despite Flexner's forty-year-old recommendation, the profession still had few, if any, journals devoted to research, few research-oriented doctoral programs, and virtually no research establishment.

In sum, across the vast literature that developed about professionalization, we can detect two umbrellalike themes (see also Wilensky 1964). The first, the one that isn't especially problematic, is the ideal of service (Goode 1969). This consists of a commitment on the part of professionals not to abuse their knowledge and skill, to serve and not to exploit their clients, and to work for the greater community good. This commitment is often embodied in a code of ethics; belief in it allows the community to grant the profession a monopoly on practice, the right to govern itself, and the authority to develop and regulate the content and format of professional education. Social work appears to have achieved this ideal.

Both Flexner and Greenwood—and even Abbott from a different perspective—understand that to gain community sanction, a profession needs a second broad attribute, a body of knowledge not available to laypersons, that confers authority, guides service to clients, and forms the content of professional education. Since professions claim unique status as the overseers of an abstract body of knowledge, they must ensure that the material taught in professional schools is sufficiently esoteric and sufficiently complex that their claims will be respected. Recall that prior to the last century, professional practice remained relatively primitive. Medicine consisted of prayer and placebo; law involved the management of small quarrels; the ministry identified and purged minor sins; and social work offered moral uplift to the wayward.

The mastery of a body of abstract knowledge and theory, rather than of practice skills per se, is viewed as the criterion that distinguishes professions from other occupations. That is why professional training is no longer solely practical, gained in apprenticeships, but has shifted from the field to professional schools affiliated with universities. Many occupations require the mastery of substantial knowledge and the development of certain skills, but only professions claim that their new members can appreciate, understand, and participate in the broad knowledge-building process underlying practice. The image of the professional presented to the public is that of a scholar-servant immersed in learning but dedicated to service.

As professional education migrated into universities, a division arose between theoretical and practical learning and between those who were primarily involved in the development and refinement of professional knowledge and those who were practitioners. While the introduction of science and scholarship into education helped validate social work's claim to professional status, it also ushered in a new structural problem, namely, how to connect the methodology of science and the abstract world of theory to the practitioner's helping specific clients with their idiosyncratic human troubles. The struggle to make this connection and to overcome the division of labor between researchers and practitioners is the focus of this book.

Social work educators know that developing and disseminating knowledge to trainees is not an assembly-line series of tasks. The enterprise is full of ambiguities and uncertainties. There are many complexities to culling usable knowledge from other scientific disciplines; developing, testing, and refining specific knowledge components using scientific methods; and transmitting this knowledge to students who are supposed to not only learn it but also remember and use it as they begin practice. Moreover, once in the world of practice, professionals are expected to stay current with developments in their specialties, continuously updating their knowledge and skills so that their clients benefit from the latest professional and scientific advances. This implies that professionals must have the requisite research skills to critically assess scientific reports and new knowledge claims and the motivation to do so.

All this is, of course, highly idealistic. Borrowing knowledge from other disciplines can be time-consuming and difficult, even for professors engaged in such work full time; developing usable knowledge and testing it for applicability and effectiveness is fraught with methodological problems; teaching professional knowledge and skill is closer at times to art than science; and devising effective methods of keeping practitioners abreast of the latest information is a constant challenge. In later chapters we review some of these efforts.

Despite Greenwood's assurances in 1957 that social work had become a profession, there were continuing disputes about whether social work indeed had an identifiable knowledge base. And as if this were not enough, some social work scholars raised epistemological questions about whether the profession was using appropriate forms of scientific inquiry.

The Problematic Knowledge Base of Social Work

Ironically, a few years after Greenwood declared that social work was indeed a profession, the newly founded National Association of Social Workers (NASW) organized a three-day conference for a highly selective group of practitioners and scholars to consider the topic of "social work knowledge." Among them were Harriett Barlett, William Gordon, Alfred Kahn, Alfred Kadushin, Henry Maas, Edwin Thomas, Bertram Beck, Chauncey Alexander, Genevieve Carter, and Katherine Kendall—all prominent leaders who continued in important roles in the profession for many years. The people invited to that working conference, which was held in November 1962 in Princeton, New Jersey, had all been actively concerned with the conceptualization of social work knowledge. They were asked to consider (in their background papers and discussion) how that knowledge could be identified, selected, assembled, and organized for the profession. This was one of several efforts that NASW undertook to define the purpose and nature of social work practice.

The report of the conference, published two years later (NASW 1964), opens with a startling sentence: "Social work has not produced a systematic body of knowledge, although it exhibits many of the characteristics of a profession" (iii). It recognizes that social work practice and education had until recently been fragmented into specialties and different organizations and lacked "frames of reference and systematically organized theoretical propositions for bringing order into its thinking" (iii). Although creative scholars were making individual contributions, they were not collectively producing a cumulative body of knowledge.

The reasons were several: social work addressed a vast array of human problems; the profession was more concerned with "doing and feeling" than with "analyzing"; and research had remained separate from practice—and researchers separate from practitioners (iv). Social work had focused more on identifying its values than on building its knowledge; it rested too much on "conviction" and not enough on a theoretically organized body of research results and practice experience (v). The conferees state that "the need for building social work knowledge becomes increasing urgent" (vi) and that "the urgency of action is increased by the need to develop research in social work practice. Theory-building in social work will require constant testing and validation in practice" (v). Thus, in their view, professional status required a knowledge base as a foundation for interventive skills.[2] And increasingly, "knowledge" referred to principles derived from some form of scientific inquiry, not merely based on conviction and values. Thus, using scientific methods became integral to the profession-building efforts of social work. The need for more research on and about practice and for encouraging connections between researchers and practitioners became a mantra that would be heard for the next half century.

Although the simple question of whether social work was a profession disappeared slowly as credentialing and licensing boards acknowledged individuals' expertise, issues about the appropriate methods for developing social work knowledge unexpectedly erupted in a contentious debate in the 1980s (for a helpful review, see Reamer [1992]). The field's reliance on traditional social science as a means of advancing its knowledge has always been a subject of controversy (see chapter 2); however, this new debate involved the philosophical foundations of the social sciences and social work research.

The Epistemological Controversy

A number of critics began to question those foundations (Haworth 1984; Heineman 1981; Tyson 1992; Witkin 1989), based on the inadequacy of the epistemological paradigm commonly taught, used, and accepted by social work researchers. This paradigm, they said, was based on the tenets of a school of philosophy—logical positivism—that no longer had any credibil-

---

[2] This point was made fifty years earlier by Porter Lee from the New York School of Philanthropy, who spoke just after Flexner and Frankfurter. Lee argued that professional status implied expertness based on scientific understanding and that social work needed to establish the technical foundations and processes of casework and social investigation upon which professional skills would rest.

ity among philosophers of science. More specifically, conventional research methodology placed undue value on quantitative approaches, experimental designs, objective measurement, and statistical analysis, borrowing from the physical sciences methodologies often ill suited for study of the ever-changing and elusive complexities of social phenomena. Similar criticisms were expressed about research in related professions (Fishman, Rotgers, and Franks 1988; Gergen 1985; Guba 1990). In social work, the critics proposed new paradigms, including constructionism (Witkin 1990) and the heuristic approach (Heineman-Pieper 1989), which, they contended, provided more suitable frameworks for the development of social work knowledge.

These criticisms and proposals were challenged by a number of researchers (Fraser, Taylor, Jackson, and O'Jack 1991; Geismar 1982; Hudson 1982b; Schuerman 1982). Counterarguments by these and other authors rejected the notions that social work research was an offspring of logical positivism and that it applied the methods of hard science inappropriately. Moreover, they pointed out, the proposed new paradigms had not yet resulted in any appreciable knowledge development.

During the 1990s, the debate took a different form. Although radically different views about the epistemological bases of social work knowledge continued to be expressed, there were fewer acrimonious exchanges in the scholarly journals. Efforts were made to create a "selective integration of epistemological perspectives," as Reamer (1992) put it (e.g., Harrison 1994; Peile 1988), but they have remained in the wings while two opposing confederations have emerged. One is a collection of epistemologies referred to under the umbrella term of "postmodernism." The other is a group of perspectives sometimes referred to as "postpositivism" (Phillips 1987).

Social constructionism has emerged as the dominant epistemology in postmodern thought in social work (Witkin 1991, 1999). A central tenet of this position is that knowledge of reality is constructed through language and human discourse. As Dewees has put it, "Realities, or beliefs, are constituted through language that creates or perpetuates shared meanings . . . there are no objective or essential truths" (1999:33). Furthermore, "There is no intrinsic reason, apart from the interests of particular groups, to privilege one form of writing and speaking or to limit knowledge claims to certain criteria" (Witkin 1999:7). Since there is no way to determine ultimate truths about reality, "scientific beliefs are products of their times" (McQuaide 1999:412). The goal of progressively building a scientific knowledge base is therefore rejected.

It is helpful in this discussion to distinguish between two forms of contructionism, which Gross and Levitt (1994) have labeled "weak" and "strong." The weak form refers to how people construe reality—concepts, perspectives, and world views are all constructions, which are of considerable interest in the social sciences as well as in the human services and have long been studied through traditional (mainstream) research methods. The strong (or "radical") form consists of the position advanced by Dewees, Witkin, and McQuaide quoted above: what passes for truth or objective reality are simply constructions produced by human language. Constructionism in this form downgrades the importance of mainstream research in social work and related fields: it may be of interest in presenting a certain kind of discourse but is no longer the ultimate means of acquiring the best knowledge possible. Gross and Levitt have expressed well how science is viewed in this form of constructionism: it

> is not . . . a body of knowledge and testable conjecture concerning the "real" world. It is a *discourse* devised by and for one "interpretative community." . . . Thus orthodox science is but one discursive community among the many that now exist. . . . Consequently its truth claims are irreducibly self-referential, in that they can be upheld only by appeal to the standards that define the "scientific community" and distinguish it from other social foundations.   (1994:43)

Not surprisingly, most of the social work practice literature using the constructionist framework pretty much ignores empirical research. Constructionist research itself, which calls for qualitative inquiry, has been quite limited. It remains to be seen what might be stimulated by recent methodological work, including a "constructivist" text on qualitative research methods (Rodwell 1998). (The terms "constructionist" and "constructivist," while having somewhat different meanings according to some writers, tend to be used more or less interchangeably in the social work literature; we use the former term and its variants.) It also remains to be seen what distinctions emerge between constructionist research and other kinds of qualitative research carried out within mainstream frameworks.

Much of the burgeoning clinical literature that labels itself constructionist can be viewed as relying essentially on the weaker form of the notion, even though it may use the language of the strong form (usually in the opening paragraphs). For example, in introducing their "two-story technique" for helping clients "open space," Hill and Scanlon make the (strong form) constructionist point that "there exists no valid and reliable way to determine the truth or accuracy or 'realness' of one person's descriptions of 'reality' over another's" (1998:76). However, their technique simply involves helping clients give descriptive names to their experiences and then discuss them in ways that enable them to gain awareness of other aspects of their realities. It could be fitted into any number of intervention approaches and into virtually any epistemology. The technique could, of course, be seen as "constructionist," but not in any meaningful epistemological sense.

We use the term "postpositivist" to refer to developments in the philosophy of science supportive of mainstream research since the decline of logical positivism, including the work of such philosophers as Bhaskar (1975); Bunge 1996), Dewey (1938), Kitchner (1993), Lakatos (1972), Nagel (1997), Phillips (1987, 1992), Popper (1959), and Siegel (1987). In contrast to constructionism, this body of work has posited the existence of an objective reality that is knowable, however imperfectly, to outside observers. (Postpostivists, as we use the term, are epistemological realists.) Science is humankind's most powerful means of attaining the best knowledge possible about this reality, which does not deny the validity of other forms of knowledge. Criteria can be developed to determine what constitutes a valid claim to truth. Scientific knowledge is cumulative and progressive—we have more useful knowledge today than we did a generation ago and will know more a generation hence. Such principles form the framework used in our appraisals of developments in social work research and knowledge, which make up much of this volume. In the section to follow we focus upon a core component of this framework—considerations in appraising knowledge—and try to show how these considerations are used in both social work research and practice.

APPRAISING KNOWLEDGE

How is the profession of social work to evaluate the truth claims of knowledge? In our everyday lives, we are constantly sorting out streams of information into what we regard as true, probably true, probably false, false, and uncertain. In this process we inevitably employ evaluative criteria. To be sure, these criteria are not fixed; they are fallible, correctable, and variously interpreted.

If a multi-epistemological perspective is adopted (Harrison 1994; Reamer 1992), standards are needed to evaluate knowledge claims emanating from diverse and conflicting epistemologies. Although it is not possible to devise criteria that would be universally endorsed, it may be possible to propose some that the majority of social work practitioners, scholars, and researchers would find acceptable. A basis for the development of such shared standards may be found in the notions of truth, corroboration, bias, theory testing, and generalization.

Truth. The idea of truth, however it is expressed, is central to the appraisal of knowledge. From a realist point of view, truth involves the relation between assertions about reality and the characteristics of that reality (Aronson, Harre, and Way 1995). In research, concern with truth is expressed in the notion of "validity." If instruments or results lack validity, we assume they do not capture reality and we do not take them seriously. As we all know, truth is neither readily defined nor always readily discerned. We agree with Phillips (1992) that is it is best seen as a "regulative ideal," a goal that we strive for but do not always attain. We must often settle for likely, approximate, or partial truths.

Although standards of truth are difficult to articulate and open to criticism, they are very much a part of professional life. For example, consider a principle of practice for working with families of delinquents formulated by Wood: "Family relationships are supported, not eroded—unless there is clear *evidence* that these relationships are fundamentally destructive" (Wood 1990:313; emphasis in the original). Although values are involved in saying that destructive relationships are bad, such a statement assumes that destructiveness in family relationships can be adequately defined and that evidence about its occurrence can be obtained. Wood's principle assumes that this evidence will be objective in the sense of being free of such biases as the practitioner's dislike of the family.

How does such "objective evidence" fit into a constructionist position? According to Dewees (1999:33), "common understandings or beliefs about the nature of the world and social relationships are not based on empirical or objective evidence; that is, the common agreements about how the world 'is' are generated, negotiated, and re-negotiated through social interchanges over time." Or as Guba and Lincoln have commented, "Phenomena do not converge into a single form, a single 'truth' but diverge into many forms, multiple 'truths'" (1982:13).

Few would quarrel with the idea that multiple perspectives on a situation are always possible and differences among them can certainly be negotiated. If they cannot, we are left with conflicting truth claims, only one of which can be valid (Haack 1996). In the illustrative case, the social worker would assume there is some single truth that objectively describes the nature of the family's interaction and can be judged against some standard of "destructiveness." He or she would normally gather evidence for a substantially true picture of the interaction. Subsequently this evidence might be presented to a judge who would expect it to be objective and reveal the truth.

The notion of objective truth, at least as a goal to strive for, appears to guide practical decisions in the world of social work. Practitioners gather data to determine the extent of a client's depression or of street crime in a neighborhood. Educators gather data on students' performance in the field and use it as the basis for presumably unbiased evaluations. The practitioners and educators in these examples assume that their purpose is to arrive at some approximation of the truth. Although much is written about scholars' epistemologies, we know comparatively little about notions of bias, error, mistakes, and truth—the actual epistemologies used—in the ordinary practice of social work. These might provide an interesting challenge to formal epistemologies that eschew such notions as truth and objectivity.

Corroboration. How then do we determine if a presumed knowledge claim is in fact true? One of our most common means is through corroboration. Pepper (1970) distinguishes between two types of corroborative evidence. "Multiplicative corroboration" involves person-to-person agreement, what is sometimes referred to as intersubjectivity. Do observers concur that a belief is warranted by the evidence? Do they concur that it has, in Dewey's (1938) terms, "warranted assertibility?" "Structural corroboration" requires a logical convergence of factual evidence. For example, accepting as knowledge the assertion that prenatal cocaine exposure can lead to behavior problems in children might require consistent agreement among clinicians and researchers that the evidence points in this direction. It might also demand evidence, or "background knowledge" (Bunge 1996), delineating the mechanisms by which prenatal cocaine exposure might lead to behavior problems.

An important extension of the nature of corroboration is the principle of "multiplism" (Cook 1987), which makes use of multiple methods and perspectives in knowledge building. Within a given study, several measures and modes of analysis are used to cross-validate findings. Similarly, multiple studies can be focused on the same research question. No single design, method of measurement, or analytic technique is seen as inherently superior. For example, experiments and quantitative analysis have their strengths and limitations and are better suited to certain questions than others. What is needed is less reliance on any particular methodology and more use of methods in combination so that the weaknesses of one are offset by the strengths of another.

But all of these forms of corroboration assume real-world referents, that is, that different observers are evaluating what is actually there and not simply comparing "constructions." Moreover, standards exist indicating what kind of evidence may be more credible than other kinds. Thus there may be a high degree of intersubjective agreement among prosecutors, witnesses, and the jury that a defendant is guilty of murder, but the verdict may be reversed in the light of DNA evidence proving that someone else committed the crime. Such evidence is regarded as providing an incontrovertible depiction of a critical piece of reality relating to unique characteristics of the perpetrator and thus overrides other evidence, even eyewitness testimony, more likely to be off the mark.

Bias. Bias refers to factors that can interfere with the appraisal of the truth claims of knowledge. Because standards for appraising truth are seldom clear, the presence of bias is difficult to determine. Yet the importance attached to it has been traditionally based on the assumption that there is an objective truth to be striven for. Bias's threat to the validity of social work knowledge has always been recognized, along with the desirability of identifying and controlling it. For example, in *Social Diagnosis* (1917), Richmond discusses the importance of recognizing and correcting numerous sources of bias, such as false analogy, that might interfere with drawing valid inferences from case evidence. In more recent times, practitioners have been taught the importance of awareness of misperceptions of clients from countertransference or from racial, ethnic, gender, or cultural stereotyping. "Bias" used in a broad sense includes not only perceptual bias but also bias in sampling, measurement, interpretation, and so on. Bias becomes virtually tantamount to error. Thus control of it is necessary in efforts to reveal the truth.

Theory Testing. Much social work knowledge is organized into theories—systems of concepts and hypotheses designed to explain and predict phenomena. Thus we have assessment theories that attempt to explain depression, delinquency, and other human problems and practice theories that attempt to show that certain interventions will change them. Theories are systematically evaluated in research through tests that either confirm or refute their hypotheses. The latter is especially important. As Popper (1959) has argued, a proposition that is truly scientific is one that can refuted or falsified. The tests themselves, and their interpretations, follow procedures carefully designed to separate truth from error. For example, to have confidence that he or she has demonstrated that two variables are related in a predicted way, the researcher must examine and rule out various sources of possible error—chance factors, instrument bias, and so on.

Although a theory may survive a number of failed tests, it cannot prevail indefinitely if empirical support is lacking. This notion has been well developed by Lakatos (1972), who has suggested specific criteria for appraising the long-term success of a theory over a course of repeated testing. Although it may still be difficult to refute a theory (Hesse 1980), it is quite possible to decide between a theory and its rival. For example, suppose a theory repeatedly predicts x but x does not occur. Furthermore, the theory is not capable of predicting y and z. A new theory proves capable of predicting not only x but also y and z, and at the same time can explain the successful predictions of the older theory. With such an outcome, the newer theory presents itself as a rational choice. The process takes time. As Lakatos (1972) has said, there is no "instant rationality"; no "crucial experiments" can decide the issue right now and once and for all.

A good example of the application of these ideas to social work knowledge can be found in the rise and fall of the "double-bind" theory of schizophrenia, originally proposed by Bateson, Jackson, Haley, and Weakland (1956). To put it simply, the double-bind theory posited that paradoxical and confusing communication sequences between parents and offspring were influential in the subsequent development of schizophrenia in the offspring. The theory attracted considerable interest, and a substantial literature on double-bind communication soon developed (Watzlawick, Beavin, and Jackson 1967). Hypotheses generated by the theory were tested in a number of studies and generally failed; moreover, serious difficulties were encountered in operationalizing key concepts (Olson 1972). As a result, the theory gradually "lost ground," as Lakatos (1972) would put it, to rival explanations of the etiology of schizophrenia. This process took more than a decade.

A rival theory with greater empirical support posited biological origins for schizophrenia but also hypothesized that the course of the disease could be affected by family factors, such as parental expressions of criticism and hostility toward the schizophrenic (Anderson, Reiss, and Hogarty 1986). This theory and related intervention approaches have received considerable support from research conducted over the past decade (for reviews, see Penn and Meuser 1996). However, like any set of ideas based on research, the theory may be eventually displaced by rivals that have not yet appeared on the horizon. We should in fact hope this will be the case. The rival will win out by better accounting for the empirical realities discovered by its predecessor and adding new ones. Regardless, the realities about schizophrenia discovered up to this point represent additions to prior knowledge and lay the groundwork for further knowledge. Thus, the knowledge produced by science is accumulative and progressive (Kitchner 1993), not simply the product of its time.

Generalization. Knowledge development for the profession emphasizes propositions that are generally applicable, despite the importance of case-specific and other forms of ideographic knowledge. Indeed, general knowledge pertaining to such matters as the dynamics of human development, the nature of psychosocial problems, or the principles of intervention provides guidance for work with specific situations. In common language, we may say that a proposition is generally true. In research, we may say that it has "external validity": it is likely to hold true beyond the limits of the study at hand.

Social work practitioners and researchers usually build generalizations through a logical process (Fortune and Reid 1999), by extending what has been learned from a given situation to others similar to it while keeping in mind how the situations differ. Generalizations are extended through convergence of findings from repeated studies, following the principles of multiplism discussed earlier. The process yields at best tentative propositions—working hypotheses, not conclusions (Cronbach 1975). Such hypotheses suggest what is likely or possible in a given situation, but whether it occurs must be determined in the new situation itself.

This view of generalization is not out of keeping with certain constructionist notions. For example, most social work researchers would be comfortable with Lincoln and Guba's principle that the "*transferability* [of findings] is a direct function of the similarity between . . . contexts. If Context A and Context B are sufficiently congruent then the working hypotheses from the sending or originating context *may* be applicable in the receiving context" (1985:124). Few propositions hold across all situations. To say that a program has been found to be "effective" is to say very little without specifying what the program consisted of, for whom it made a difference, under what conditions, and so on.

The knowledge base of social work is ill defined and difficult to identify, delimit, or organize. Moreover, most of it is not the product of rigorous scientific testing. Although there are different views about the appropriate forms of scientific inquiry in social work and the standards that should be applied in assessing knowledge claims, "fuzzy" notions of truth, corroboration, bias, theory validity, and generalization contain useful criteria for evaluating knowledge. We may disagree on how particular criteria are to be conceptualized, defined, or applied in given cases, but they provide the basis for appraising knowledge claims in social work. As Rorty (1979:316) has said, "The dominating notion of epistemology is that to be rational . . . we need to find agreement with other human beings. To construct an epistemology is to find the maximum amount of common ground with others." Disputes at the grand epistemological level are likely to have few practical consequences, and for social work it is the practical consequences that matter.

Overview of This Book

At least since Flexner's proclamation that social work had not yet achieved professional status, there has been continuing discussion about the proper role of knowledge, science, and research in the practice of social work. This book is an examination of significant elements in the evolution of that discussion. From the days of scientific charity, the profession has struggled to find a comfortable, practical, and effective method of using systematic inquiry to develop, direct, improve, and evaluate practice. Over the years, social workers and scholars have embraced and promoted many different methods of bringing science to bear on practice.

Two different, overarching methods or strategies can be discerned. One has been to make practice itself more like research, to have it mimic scientific inquiry by emphasizing systematic gathering of information, careful study of individual clients, and decision making about intervention based on the analysis of case data. There is a common thread between using research techniques as guides for practice in scientific charity to using single-subject designs emphasized by advocates of the scientist-practitioner model. A second general strategy has emphasized practitioners' use of the results of scientific inquiry in their work with clients. This can be seen in social workers' attempts to harness the findings of large-scale social surveys or social experiments, the rapidly expanding general social science knowledge (primarily from psychology and sociology), and the burgeoning medical sciences. From the early attempts to forge links between social science and social work to the promotion of research utilization to the more recent emphasis on empirically supported practice, there is a common emphasis on practitioners not doing research but instead being guided in their practice by the fruits of others' research.

Each of these two general strategies has repeatedly found expression in particular sociohistorical circumstances that framed the problems of clients and the externalities of the developing profession. These organized attempts to connect science and social work have often been controversial and have had opponents who questioned their philosophy, practicality, or politics.

In this book, we offer a critical appraisal of the strategies and methods that have been used to develop knowledge for social work practice. Our method will be to identify the major approaches, placing each one in historical perspective by explaining the nature of the problems that it attempted to solve. We offer a balanced appraisal of the promises, accomplishments, and limits of these efforts. Although our emphasis is on face-to-face work with individuals, families, and groups, the ideas we advance may have relevance to less direct forms of practice, such as administration and community organization.

In chapter 1 we have indicated the central role that knowledge plays in the world of professions and tracked how concerns about social work's professional status led to concerns about its knowledge base. Systematic inquiry is fundamental in developing, refining, and testing practice knowledge, so concerns about knowledge generated concerns about the research enterprise. Those have been enlivened by epistemological issues regarding the proper validation of knowledge claims. We have suggested one approach to assessing such claims.

In chapter 2, we provide a broad historical overview of the attempts to create a role for science in the profession up to the mid-twentieth century. Science served as both a model and a source of knowledge for practice. The chapter begins with a description of the early uses of the scientific method as a model for casework practice, from Mary Richmond to the psychoanalytic movement. We discuss two types of knowledge—assessment and intervention—in both their case-specific and general forms. With regard to the latter, we review the history of studies that addressed the effectiveness of social work and those that focused on the process of intervention. Finally, we review the slow development of a research infrastructure. The failure of early experiments to demonstrate the effectiveness of intervention and the rise of demands for accountability within social welfare programs produced a crisis in casework in the early 1970s that reverberated throughout the profession, particularly in academic settings. At the same time, a broad alteration in the infrastructure and context of research in social work was taking place, involving an explosion in the growth of doctoral programs and the training of a new cadre of social work researchers who abandoned the "soft" science of psychoanalytic inquiry to embrace more behaviorally oriented theories, experimental designs, and quantitative techniques. Social work research moved from its traditional base in the agency to the academy.

In chapter 3, we review the mutual interests of practitioners and researchers in organizing practice knowledge in terms of clients' problems. We revisit how problem diagnosis and classification have evolved and how current classification systems have departed from earlier traditions. We then evaluate a contemporary attempt to de-emphasize problems as the organizing foci of intervention knowledge.

In chapter 4, we examine a major change in scientifically based practice that began in the 1960s. As it evolved, the effort called for practice to be scientific in the sense of being a rational, systematic, problem-solving activity. The promotion of the scientist-practitioner, who would be trained to use single-system designs and a handful of research techniques, was a response both to the call for accountability and to the recognition that practitioners were not reading research reports. Borrowing from clinical psychology, an influential group of research-oriented professors set out to revise not only the nature of practice education but the nature of social work practice itself. This chapter reviews the history of the scientist-practitioner movement, describing its origins, characteristics, and promise. Although unsuccessful in its most ambitious objectives, it nevertheless stimulated wide-ranging changes in the structure of graduate programs and vigorous debates about the merits of making practitioners into clinical researchers.

There is no simple method of connecting science to practice, although we witness obvious links when we use airplanes, computers, TVs, or medications. We are surrounded by commercial products that originated in scientific labs but through engineering and marketing have made their way into our hands and homes. In industry, the process of turning basic scientific discoveries into usable products involves the work of intermediaries—engineers, product designers, marketing specialists—and a series of research and development processes. This observation prompted several social work scholars (most prominently, Jack Rothman and Edwin Thomas) to examine whether it was possible to develop R&D procedures for social intervention and thus to solve both the problem of ineffective methods and the problem of dissemination and utilization. In chapter 5, we review these attempts to link research and practice in a complex, sequential process and explore why, given their promise, they have not yet enjoyed the success of industrial R&D.

It is hardly surprising that social work researchers looked to industry and engineering for a way to forge links between the research and practice worlds. Equally predictable was that they would look to computer technology for assistance with professional tasks. Both research and practice involve collecting and manipulating complex sets of information. Computers have made this increasingly simple and efficient. In chapter 6, we examine how the developments in technology over the last fifty years were envisioned to assist practitioners and improve the delivery of social services. Early applications such as management information systems and expert systems were viewed as new methods of linking science and practice. Many more such links are emerging, each with its own promise.

Recently, scientific knowledge has been brought to practitioners in research-supported practice. Social workers use methods whose effectiveness has been demonstrated through empirical research, developing intervention protocols from successful experiments. In theory at least, practitioners use such protocols, written treatment instructions, or guidelines in implementing the interventions. Major initiatives to identify effective methods and put them in the hands of practitioners are taking place in psychology and medicine. Similar efforts have also appeared in social work and are likely to intensify. Although it may be an important means of putting social work on firmer scientific ground, empirical practice raises a number of issues. What criteria should be used to determine which interventions are efficacious? Is there enough research of sufficient rigor on which to base an adequate repertory of interventions? How do we deal with threats to effectiveness when practitioners depart from established guidelines or apply procedures to populations other than those used to validate the interventions? In chapter 7, we discuss the origin and character of empirically based practice, its goals, the issues to be resolved, and whether it is likely to succeed in social work.

In chapter 8, we review one of the ways in which social work has tried to incorporate knowledge into practice—through the dissemination and utilization of research by practitioners. After tracing the rise of such concerns, we examine how the profession attempted to prepare students to consume research. It soon became apparent that practitioners, in both social work and other professions, did not necessarily pay attention to the research journals. Knowledge generation did not automatically or efficiently lead to knowledge dissemination or utilization, and some scholars at first tried to pin this on practitioners' negative attitudes and agencies' self-protective nature. The Council on Social Work Education and the National Association of Social Workers joined scholars from other disciplines in discussing what could be done to enhance the utilization of research. This chapter reviews the emergence of these concerns and their evolution from the simple notion that practitioners should be taught to "consume" research to the recognition that the dissemination and use of knowledge takes place in a much more diffuse, uncertain, and multilayered manner.

In the final chapter, we reflect on what we can learn from the various attempts to develop knowledge for social work practice. We recognize how difficult it has been to bring science to practice, but argue that we have made progress—hesitantly and unevenly to be sure—in understanding how to improve the knowledge base and in recognizing the kind of sustained efforts that will continue to be needed. Future progress will require more realistic expectations about what is practical and better efforts to develop considerably more robust collaborative structures that bridge the gaps between practice and research. This will remain a serious challenge, because the "soft" nature of social work knowledge impedes cumulative knowledge-building efforts. The graduate schools of social work in major research universities are likely to bear the burden of creating these structures, for there are no other institutions available to fully address the concerns first raised by Flexner and Greenwood.

CHAPTER TWO

Science and Social Work:
A Historical Perspective

There are two major ways in which professions have made use of science. One is by following a scientific model in conducting professional activities. The other is by using scientific knowledge to inform those activities. This depends on the development of an infrastructure for generating such knowledge—a cadre of researchers, financial resources, training programs, methodological texts, organizational settings, and so forth. In this chapter we shall trace the evolution of these uses of science in social work and the development of a supporting infrastructure, beginning in the latter part of the nineteenth century and continuing somewhat beyond the midpoint of the twentieth century.

Science as a Model for Practice

In the early decades of its development, social work used science mainly as a model for practice, beginning toward the end of the nineteenth century with the Charity Organization Society (COS) movement, which had begun in London in 1869. By the 1890s almost a hundred COSs were operating in American cities (Watson 1922). Their purpose was twofold: to coordinate the work of existing relief agencies by registering the poor in need of assistance, evaluating their needs, and making appropriate referrals; and to provide guidance to the poor through the "friendly visiting" of volunteers and paid agents. The COS was not designed to give tangible assistance itself but rather to systematize relief giving and, through education and moral uplift, work with the poor to obviate the need for material help.

Leaders of COSs, as well as educated people generally, were impressed by the triumphs of science in such fields as medicine, biology, and engineering. The leaders were quick to claim that the scientific method could be applied to their work. Josephine Shaw Lowell, one of the founders of the New York Charity Organization Society, minced no words: "The task of dealing with the poor and degraded has become a science and has its well defined principles recognized and conformed to closely by all who really give time and thought to the subject" (1884:Preface). In remarks at a meeting of the American Social Science Association in 1880, D. O. Kellogg, director of the Baltimore Charity Organization Society, declared, "Charity is a science, a science of social therapeutics, and has its laws like any science" (Germain 1970:9). As Charles S. Loch, director of the London Charity Organization Society, succinctly observed, charity work is "not antagonistic to science: it is science" (1899:11).

Lowell, Kellogg, and Loch were advocates of what would be called the "scientific charity movement." Charity was to be, like science, an orderly, systematic process in which practitioners gathered facts, made hypotheses, and revised them in the light of additional facts from the case (Evans 1889). The volunteer "friendly visitor" or paid agent of a charity organization was to gather relevant facts bearing upon the plight of the individual or family seeking assistance and develop them into hypotheses that might provide causal explanations. For example, what might be the causes of the person's indigence? Inability to work because of poor health? Failure to keep jobs because of drinking? The hypotheses could then be "tested" with further evidence from the case. Suppose drinking on the job appeared to be the problem. Perhaps this could be confirmed through a contact with the employer. Well-tested hypotheses could explain the problem and suggest possible remedies. The methodical and systematic approach of the physician applying the scientific art of medical diagnosis was taken as a model. This was, of course, a borrowing of the methods of science. The scientific knowledge that might have informed charity efforts was just beginning to appear in the form of studies of the poor.

The notion of using a scientific approach in work with the indigent fit well into the larger rationale of the emerging charity organizations. They wanted to give material assistance and guidance to the poor in an efficient, businesslike manner. Only in this way, the argument went, could the organizations cope effectively with the growing problem of urban poverty. As the epitome of rational problem solving, a scientific approach could help ensure that methods of dealing with the poor would be systematic and effective.

Also, to claim that activities constituted a "science" was a way to establish credibility and to assert that the activities demanded special knowledge and skills. As social work functions became more the province of paid agents, this claim became a way of achieving professional status. This was especially important to social workers, who wanted to differentiate themselves from informal helpers who gave material assistance and guidance to the needy (Robinson 1921).

How well this scientific paradigm was carried out in practice is open to conjecture. The case records of the time were not written to gauge the charity worker's adherence to a scientific practice model, and to our knowledge they have not been systematically studied for that purpose. In all likelihood there was a large gap between the rhetoric of the scientific charity movement and actual charity work. As Lubove (1965) has commented, "scientific social work remained an elusive ideal rather remote from reality" (20). Even if the model were applied, it would suffer from lack of scientific theory and concepts. Little had been developed in the way of an implementing methodology. For example, although Lowell (1884) wrote of the "well-defined principles" of the new science of charity in the preface to her book, the book itself contains no references to any such principles—or to scarcely anything else, for that matter, that could be identified as scientific.

Although its influence on actual practice may have been limited, the scientific charity movement established the *potential* utility of the scientific method as a framework for individualized social work practice. It set forth a rough design that successors would elaborate upon and ultimately implement. Moreover, as Bremner (1956) has pointed out, the individualized investigative approach of the visitors and agents helped develop a more objective view of the poor and their circumstances. Stereotypes of them as "shiftless" or "depraved" were challenged by factual and detailed descriptions. Scientific charity also helped establish the case record as an instrument for guiding practice and a method of collecting data for future research.

The scientific charity movement also provoked a reaction against the use of science in social work. The movement was derided by John Boyle O'Reilly, who referred to offering charity in the name of a "cold statistical Christ" (cited in Sheffield 1937:275). Skepticism about the utility of science took many forms, from disbelief that the results of casework could ever be measured (Rich 1926) to rejection of the notion of a scientific base for practice (Taft 1937). As we have noted in chapter 1, the theme is still very much with us.

The application of the scientific method to the study and treatment of individual cases was extended by Mary Richmond (1917) in *Social Diagnosis*. The profession had begun to form. From its nineteenth-century roots in work with the indigent and dependent children, social work had become established in general hospitals, psychiatric facilities, juvenile court settings, and the schools. Professional training of social workers had started. Although most were employed as caseworkers working directly with individuals and families, a definitive text on casework practice had not yet appeared. In *Social Diagnosis*, Richmond sought to meet part of this need. As the title suggests, she concentrated on the diagnostic or assessment phase of practice. Although she did not draw explicitly upon the writings of the scientific philanthropists, she constructed a model that preserved the essence of their approach: the thorough gathering of relevant facts then used to form and test hypotheses about the case. As Germain (1970) has noted, Richmond basically applied a medical model to the client's social difficulties and to social evidence bearing upon them. As she developed her model, especially in relation to obtaining and evaluating evidence, she drew upon other disciplines, notably law, history, and logic.

Richmond did not explicitly present *Social Diagnosis* as a "scientific" approach to practice. Nevertheless, it provided a detailed protocol for applying the logic and thoroughness of the scientific method to social assessment. How evidence is to be acquired, assimilated, and evaluated; the rules to follow in making inferences; the use and evaluation of collateral sources; the employment of detailed questionnaires in interviews with particular client groups; and attention to the fluctuation of symptoms over time could all be seen as part of a scientific approach to diagnosis and would not be out of place in a modern text on empirical practice. To provide a graphic illustration, one chapter presents a thirty-eight-item questionnaire for supervisors to use in reviewing case records. Most of the questions would apply as well to data collected for research purposes. "Have any marked personal prejudices of the caseworker been allowed to warp the account?" "Have leading questions been asked without full knowledge of their danger?" "Have statements been sought at first hand and not through intermediaries?" And, of central importance to the scientific paradigm, "What hypotheses and inferences of the worker and or others have been accepted without the necessary testing?"

Although Richmond did not present *Social Diagnosis* as "science," others did. A few years after its publication, she was granted an honorary master's degree by Smith College for "establishing the scientific base of a new profession" (Colcord and Mann 1930:427). Later, Klein referred to *Social Diagnosis* similarly as the "formulation of a new science" (1931:67). As with the writings on scientific charity, we lack information about the actual use of the methods of *Social Diagnosis* in practice at the time. It was, however, widely used as a primary text in training caseworkers until the late 1930s (Dore 1999) and presumably did have an impact on practice.

The paradigm presented in *Social Diagnosis* raises another, more profound issue. Richmond's formulations called for practice to be scientific in the sense of science as a rational, systematic, problem-solving activity involving thorough methods of data collection, attention to the quality of evidence, effort to be objective and unbiased, hypotheses tested against facts, and so on. These are the actions of a good scientist at work, but they also describe the actions of a good lawyer, a good journalist, a good detective, or for that matter a good plumber.

The question is when a rational, systematic, problem-solving activity becomes a use of the scientific method. As Dewey once put it, "Scientific subject-matter and procedures grow out of the direct problems and methods of common sense—but enormously refine, expand, and liberate the contents and the agencies at the disposal of common sense" (1938:66). If good problem solving can be seen as common sense, how much refinement and expansion is necessary to make it "scientific?" Perhaps what we find in *Social Diagnosis* is a "proto-science" or "advanced common sense," a mode of problem solving that is shared by science and a number of disciplines and activities. It is part way on a continuum toward a more clearly scientific form of practice. Thus in social work the addition of certain conceptual and methodological refinements, such as the collection of baseline data, the use of research instruments in initial assessment and in measuring case progress and outcome, and the employment of research-based interventions defines contemporary forms of scientifically based practice, which are farther along the continuum.

Certainly Richmond was the first to fully articulate an assessment model for social work consistent with a scientific approach. She gave practice a strong push in a direction compatible with a scientific point of view. Moreover, she established a benchmark along the "common sense-to-science" continuum. Subsequent practice models can be compared with hers to determine if they are indeed "more scientific."

Even as *Social Diagnosis* reigned as the dominant text for caseworkers, an alternative view of practice was beginning to emerge. The psychoanalytic movement began to be seen by some social workers as an advance over Richmond's fact-gathering approach. Born in the 1920s, psychoanalytically oriented casework became the dominant form of practice by mid-century and continues to be a major force. Its emphasisis on the client's subjective experiences and hidden thoughts and feelings posed difficulties for practitioners who wished to be scientifically oriented. The client's inner life did not lend itself to precise definitions and hypothesis testing. But in fundamental ways the psychoanalytic approach reinforced the notion that practice should follow a scientific model. Freud probably saw himself as more a scientist than a therapist. Psychoanalysis was presented to the world as a *science* built by the case studies of Freud and his followers (Pumpian-Mindlin 1952). A good psychoanalytic practitioner was to search for the facts and to test his or her hypotheses about them as assiduously as any follower of Mary Richmond. Hollis, a leader of the psychoanalytic movement in social work, commented, "Certainly since the days of Mary Richmond we have been committed to objective examination of the facts of each case. We draw inferences from those facts. We constantly alert ourselves to sources of error" (1963:13). The problem was, of course, that the facts were likely to be elusive, fuzzy, inordinately complex, and often difficult to distinguish from fiction. Tensions were created by an obligation to be scientific on the one hand and obstacles to being so on the other. Many psychoanalytic practitioners resolved the tensions by setting aside their scientific obligations. Others were inspired by the commitment to science to search for forms of practice more compatible with it.

Science as a Source of Knowledge

The use of the scientific method as a model for practice is a precondition for a more critical use of science—applying knowledge derived from research. In the development of a scientific art, modeling practice after science generally comes first. The model is readily available, easy to grasp, and can be implemented without delay, whereas developing a useful base of knowledge relevant to practice takes time. Conversely, although a scientific manner may be a sensible way to practice, its results may be quite limited in the absence of knowledge. For example, physicians may have carefully gathered facts and tested hypotheses about patients with inexplicable symptoms of fever and chills, but it was not until the discovery that the anopheles mosquito was the carrier of the disease that real progress was made in diagnosis and control of malaria.

TYPES OF SCIENTIFIC KNOWLEDGE

In discussing the development of the knowledge base of social work, it is useful to distinguish between different kinds of knowledge. One distinction is between knowledge that is general in form as opposed to that which is case specific. Another has to do with whether the knowledge pertains to assessment or intervention processes (Fischer 1978).

Table 2.1 shows the four categories of knowledge yielded by this two-way classification. The cells contain illustrations of each type of knowledge; of course, they all interact. For example, knowledge of typical early warning (prodromal) signs of a relapse would guide inquiry with a disturbed schizophrenic client. As this and the earlier malaria example illustrate, valid general knowledge may be essential in determining what case-specific information should be collected.

TABLE 2.1 Types of Knowledge

	ASSESSMENT	INTERVENTION
Case-specific (Case may be system of any size, individual, group, community)	Facts and hypotheses about case at hand	Information about interventions used and apparent outcomes in case at hand
General	General concepts, propositions, about problems, human behavior, communities, etc.	Generalizations about intervention and their effectiveness

Case-Specific Assessment Knowledge: The Social Survey Movement. Almost all of the empirically based knowledge used by early social workers was case specific, and almost all related to assessment. *Social Diagnosis* could be seen as a protocol for generating knowledge of this kind for work with individuals and families. The most ambitious and complex use of case-specific assessment knowledge emerged at the beginning of the twentieth century in what has been referred to as the social survey movement. The movement began in 1907 with the launching of the Pittsburgh Survey, which grew out of a concern about the impact of the combination of industrialization and urbanization on the lives of city dwellers. It was an ambitious, citywide attempt to investigate actual and potential problem areas, such as working conditions in the steel industry, industrial accidents, typhoid fever, trades employing women, and child-helping institutions. As its prospectus stated, the survey's purpose was "to supply unbiased reports in each field as a basis for local action" (Kellogg 1914:479). It was assumed that exposing the shocking facts of various urban ills would goad the citizenry into programs of reform. The Pittsburgh Survey was quickly followed by many others as the movement swept across the nation. Hundreds were eventually carried out as the survey became the dominant form of research related to social work concerns. As an indicator of their importance, the leading social welfare journal of the time, *Charities and the Commons*, changed its name to *Survey*.

The kind of knowledge generated by surveys was clearly to be assessment oriented and case specific. As Zimbalist has commented, "the primary emphasis was on 'diagnosing' a given community as it was 'today' in a major, one-shot operation. The findings and conclusions might have no bearing on any other community or indeed on the same community five years hence" (1977:132).

By the 1920s, large-scale surveys of the Pittsburgh type had begun to give way to more circumscribed, special-purpose community surveys carried out by councils and other planning bodies. There is general agreement that the surveys fell well short of achieving their objectives of bringing about significant social change (MacDonald 1960; Polansky 1971; Zimbalist 1977). Polansky delivered one of the harsher judgments: "Most community surveys were initiated in the hope that once the facts were compiled and artfully presented, the local leadership would be inspired to act." But "Experience in this country, alas, was that knowledge of needs did not guarantee incentive to meet them" (1971:1102). In terms of our framework, the citywide surveys were restricted to assessment knowledge, which may have been flawed by the biases of the data gatherers as well as their insensitivity to local conditions (many were outside experts) but still revealed problems that might have warranted action. What was lacking, however, was the necessary intervention knowledge to use the exposure of these problems as a basis for reform. Relevant knowledge might have included ways to organize key players in the community to develop appropriate reform agendas. However, this criticism must be tempered, given the extraordinary difficulties in developing effective intervention follow-through for surveys, difficulties that are still very much with us today.

But like other "failures" in social work research, the omnibus survey was eventually transformed into a very useful tool. It became more specialized and more focused, more methodologically sophisticated and more sensitive to the politics of implementation. Community surveys and needs assessments of today can be traced back to the survey movement.

From another perspective, it is not surprising that the large-scale surveys fell short of the mark. The surveys were new research technology. Some of their faults—ambitiousness, diffuseness, ignorance of local conditions, lack of methodological refinement—may be attributed to the initial use of a new research paradigm. As we shall see, quite similar faults characterized a first-generation trial of another research paradigm—the field experiment.

General Assessment Knowledge. Knowledge relating to assessment in social work covers a vast, not well-charted terrain. One type concerns the various systems social workers deal with. These can be thought of as a hierarchy of open systems (Tomm 1982); like a set of Russian dolls, they can be fitted into one another in an ascending scale of complexity. For example, neurological, circulatory, and other systems make up the individual, who in turn can be part of a family system, which is a part of a community, and so on. Another type concerns target conditions—the problems or disorders that social workers try to affect. Knowledge of a target condition subsumes knowledge of one or more constituent systems and hence is at a higher integrative level. Thus, a target condition involving a family with an alcoholic member would involve knowledge of individuals and families as well as of alcoholism.

The empirical foundations of this body of knowledge have evolved slowly since the beginnings of social work. In an effort to provide underpinnings for the scientific charity movement, ambitious studies were undertaken to determine the causes of poverty. Although these studies used data from particular cities, their purpose was to lay the groundwork for an explanation of urban poverty in general. Perhaps the best known was conducted by Amos Warner, an economist. Warner (1894) surveyed 28,000 cases of relief applicants investigated by a number of COSs that used a common form. Essentially, each investigator's own attributions of causes were entered on these forms. In one sense the study was a collection of the individual opinions and biases of hundreds of investigators, leading to such statistics as the percentages of cases in various cities in which the cause of poverty was "shiftlessness and inefficiency." But it was also one of the first major efforts in the United States to systematically study a problem of concern to social workers. The study provides a sense of how more "objective" factors, such as sickness, unemployment, and alcohol abuse, contributed to the poverty of the time.

During the early years of the twentieth century, empirically based assessment knowledge available to social workers continued to come mostly from the social sciences, with additional contributions from medicine and psychiatry. However, the adequacy of the empirical base was often open to question, and there was little evidence that social workers made use of such knowledge, regardless of its soundness. Based on a study of family agency case records conducted in the late 1920s, Karpf concluded that workers relied largely on common-sense concepts and judgments. "Despite the testimony of social work literature that social workers should have ... training in the social sciences basic to social work their case records give little indication that they have or use such knowledge" (1931:353).

There were many reasons for this, including the relative dearth of scientific knowledge and the lack of emphasis on it in the training of social workers (Karpf 1931), but a fundamental reason, one that is still very much with us today, is that the knowledge usually lacked the characteristics social workers needed in assessing case situations, that is, it did not have specific relevance to problems and people at hand or a theory with explanatory and predictive propositions. Let us consider, for example, a social worker assessing a family in financial need in which the father's abuse of alcohol has prevented him from holding a job. It does not help to learn, from the Warner study, that alcohol abuse is a cause of poverty in a given percentage of cases. That fact is immediately apparent in this particular case. What is needed is knowledge that would help the practitioner pinpoint possible modifiable factors that might be causing or exacerbating the problem. It would also be helpful to have a basis for estimating the likelihood that the problem will persist. An assessment resulting from such knowledge would provide direction for an intervention plan.

As Karpf and others were bemoaning social workers' nonuse of knowledge from the sciences, the psychoanalytic movement was beginning to provide an assessment theory rich in causal hypotheses relating to the individual's inner life and childhood experiences. If the wife in the example above tended to capitulate to her drinking husband when he threatened to leave her, the practitioner could theorize that the wife had a fear of abandonment, possibly related to a childhood experience. If she became depressed, repressed rage at her husband might be the answer. And so on. But this fertile theory had a major shortcoming—its scientific base was questionable, if indeed it even could be said to have a scientific foundation at all.

As noted earlier, psychoanalysis had been proclaimed a science based on the case studies of Freud and others and was used as a model for case assessment in social work practice. However, its key constructs and hypotheses, with their references to elusive phenomena such as unconscious processes, proved difficult to test with the tools of conventional science. It could be best regarded as largely an unverified theory during the period of its ascendancy in social work. Nevertheless, psychoanalysis revealed some of the components—e.g., relevant explanatory hypotheses—that more empirically based knowledge needed to contain in order to serve the assessment needs of social workers.

By the middle of the twentieth century, the social sciences and psychiatry, spurred by new behavioral theories, more discriminating measurement techniques, and more sophisticated analytic methods, were beginning to produce substantiated knowledge useful in the assessment process (Kadushin 1959; Kammerman et al. 1973; Stein and Cloward 1958). The information came from studies of the family and small groups, of social roles, of processes of learning and communication, of cultural and social class differences, and of the development and resolution of crises. In addition, social work research began to create a small amount of assessment knowledge. For example, in their hallmark study of family agency clients, Ripple, Alexander, and Polemis (1964) found that a favorable "discomfort-hope balance" (similar to a self-efficacy belief) was a better predictor of service outcome than measures of the client's capacity to change. However, only in relatively recent years has there been a sufficient accumulation of knowledge of this kind to have a significant impact on practice.

Another use of science for assessment purposes emerged during the early years of the profession: the use of scientific concepts and frameworks to sensitize practitioners to phenomena related to clients and their problems and to systematize the assessment process. The work of scientists produces ways of viewing the world that social workers may find of value. For example, psychoanalytic investigations, despite their shortcomings, did yield some empirically validated sensitizing notions, such as the importance of unconscious motivation, that alerted social workers to the very real possibility that their clients might be driven by factors outside their awareness. Ada Sheffield (1937) demonstrated how a "scientifically minded" social worker might use the sociological notion of "the invisible environment" of community-sanctioned mores (MacIver 1931) to understand the struggles of a poor family with too many children and the unwillingness of a social agency to give birth control information.

Intervention Knowledge. Intervention knowledge includes the activities that social workers carry out in their work with clients, the outcomes or effects of these activities, factors that might influence the social workers' methods, and their interactions with clients in the course of this work. Case-specific forms include practitioner data about interventions used or outcomes in a case as well as evaluations done for agencies' own planning purposes. General forms include intervention typologies, data on the use of different kinds of interventions, published program evaluations of general interest, and experimental tests of particular intervention methods. We shall focus here on general knowledge relating to the study of social work outcomes and intervention processes, perhaps the two best known and most important types.

EARLY RESEARCH ON OUTCOMES

Although there was interest in evaluating the outcomes of social work intervention and recognition of its importance during the early decades of the profession, little research was actually done. In one of the earliest reported studies (Brandt 1906), COS workers were asked to judge clients' progress in various areas of functioning, such as "temperance" and "judgment." The first major evaluative study in social work did not appear until the early 1920s. As reported by Theis (1924), a child welfare agency in New York attempted to follow up on children placed in foster homes during its 25-year history. Data were obtained, mostly through interviews, on almost 800 individuals, then all adults. Their current adjustment was judged to be either "capable" (77 percent) or "incapable" (23 percent). The high percentage judged capable was taken as a positive evaluation of the agency's foster care program.

These early outcome studies, and others that could be cited, represented rather crude efforts at determining the effectiveness of social work. Judgments were made by people with vested interests in the programs, with little attempt to determine measurement reliability. Most important, there were no controls for the possibility that whatever gains were observed could have been caused by factors other than the intervention. The lack of rigorous research on outcomes prompted Richard C. Cabot, in his 1931 presidential address to the National Conference of Social Work, to exhort social workers "to measure, evaluate, estimate, appraise your results" (cited in MacDonald 1960:12).

The Effectiveness Crisis. Cabot's remarks were a prelude to a growing demand for studies of the effectiveness of casework programs. Funding agencies were asking for proof that such programs were producing results. Program designers and proponents wanted scientific evidence supporting their work, to be used as a basis for program justification and expansion. The profession as a whole had a strong stake in determining if the methods used by practicing professionals and taught in schools of social work were indeed effective in alleviating client problems. There was little empirically based intervention knowledge. Evaluation research was seen as a way to build it. For these reasons, there was increasing interest in using more rigorous evaluation designs using some form of equivalent control group, individuals receiving either no service or some form of lesser service.

Controlled experiments on casework programs began in the mid-1930s with the launching (by Cabot himself) of the Cambridge-Somerville Youth Study, a project designed to determine if counseling and other services for troubled youths could prevent delinquency (Powers and Witmer 1951). Controlled evaluations over the next three decades tested additional casework programs for delinquents or predelinquents (the heaviest concentration of studies), children with adjustment problems, families receiving public assistance, and the frail elderly, among other populations.

By the 1960s, as the controlled evaluations accumulated, a disconcerting trend in their findings became apparent: in most of the studies, clients receiving services of trained social workers were not showing more progress than counterparts receiving none or lesser services. For example, six evaluations of casework and group work services to predelinquents and delinquents had been conducted. None had been able to demonstrate the effects of social work intervention (Fischer 1973). In a major evaluation of a social work effort to help clients on public assistance to improve their functioning and achieve financial independence, those receiving professional social work services did little better than controls. The project, known as the Chemung County study, was conducted in rural upstate New York (Wallace 1967). A prestigious social agency in New York City, the Community Service Society, took the findings as a challenge. It designed an intervention program for a more promising client group, those who had just recently begun to receive public assistance, and tested it in cooperation with the New York City Department of Social Services. The agency assigned some of its most skilled and experienced practitioners to the project. But the results were no better than those of its Chemung County predecessor (Mullen, Chazin, and Feldstein 1972). The bad news about casework even made the national press. When *Girls at Vocational High* (Meyer, Borgatta, and Jones 1965), a large, well- designed, multiyear experiment in delinquency prevention, also produced dismal findings about the effectiveness of social work intervention, it became the subject of a column in the *New York Herald Tribune* by a well-known science writer, Earl Ubell, ominously entitled CASEWORK FAILS THE TEST.

This string of poor showings prompted a number of symposia and publications (Fischer 1973, 1976; MacDonald 1966; Mullen and Dumpson 1972; Wood 1978). In one medium or another, numerous social work scholars, researchers, and practitioners had something to say about the "effectiveness crisis" in a prolonged exchange that lasted well over a decade.

They did not speak with one voice. Some scholars, notably Fischer, who published the most influential and best-known review of the studies (Fischer 1973, 1976), took the position that the projects were valid efforts to prove the effectiveness of casework. That they were unable to do so suggested that the modes of practice examined were deficient. An overhaul of casework methods was in order. Two leading members of the social work practice community, Helen Perlman (1972) and Carol Meyer (1972), did not see the studies as a nullification of casework but as an indication that the casework programs studied were badly conceived and did not follow basic practice principles. Thus, poor results were not surprising. Others, primarily researchers, attacked the studies rather than casework in general or the specific programs (e.g., Cohen 1976; Hudson 1976; MacDonald 1966), pointing out that the research itself was so flawed that it really said little, one way or another, about effectiveness. In particular, casework itself was too poorly and variously defined in the studies to warrant any conclusions.

From our perspective, there were deficiencies in both service and research strategies. Service goals were often ambitious but without well-articulated methods of achieving them. For example, in projects involving families receiving public assistance, a principal goal was to restore economic independence. In projects involving troubled youths, a goal was to prevent delinquency. But in neither case was it clear how these objectives were to be achieved. In the public assistance projects in particular, there was no reason to believe that professional caseworkers could offer anything more potent to restore financial independence than the untrained "lesser service" workers who were already skilled in helping clients locate jobs and other resources (Meyer 1972; Perlman 1972).

Goals also tended to be diffuse—e.g., enhancing social functioning or ego strength. This led to global, imprecise outcome measures such as scales to assess family functioning or client "movement" and to possible mismatches between what caseworkers were trying to achieve and what researchers were trying to measure. For example, in *Girls at Vocational High*, goals were "described in such terms as: to increase self-understanding, to develop more adequate psychological and social functioning, to facilitate maturation, to supplement emotional resources inadequate for the ordinary and extraordinary stresses of adolescence" (Meyer, Borgatta, and Jones 1965:181). Given this diffuseness, the caseworkers might have concentrated on helping many of their clients with emotional or interpersonal problems or increasing their self-understanding without getting directly into their academic difficulties. Also, they had little contact with the girls' teachers and parents, yet changes in school grades were used across the board as an outcome criterion. Current perspectives suggest not to expect much change in an adolescent's academic functioning without a major effort concentrated on academic problems involving students, teachers, and parents (Bailey-Dempsey and Reid 1996; Reid and Bailey-Dempsey 1995). Moreover, the diffuseness of goals meant in general that caseworkers might try to achieve certain goals in only a small minority of cases, whereas all evaluation instruments were applied to all cases. Some might have involved the kinds of personality changes measured by the Junior Personality Quiz, one of the instruments used. The few cases in which success was achieved in respect to this goal might not have been sufficient to register as a statistically significant difference.

The intervention method most commonly used in the projects, psychodynamically oriented casework, appeared ill suited to the kinds of problems and clients most frequently dealt with. Its emphasis on helping clients express feelings and develop understanding might have been effective with depression and other emotional disorders but was probably not helpful in bringing about behavior change in delinquents or helping poor families achieve economic independence. For example, subsequent research has shown little support for psychoanalytically oriented approaches with delinquents (Andrews et al. 1990; Reid 1997a). Moreover, outcome measures used in the projects were not designed to capture the kind of subtle changes in feelings and perceptions that might result from such methods.

The question of fit between the clients' felt needs and the program goals and methods could be raised about most of the projects. At the time they were conducted and increasingly since, social workers have subscribed to the notion that successful efforts to bring about change must involve partnerships between social workers and clients. For most problems dealt with in the projects, for example, delinquent behavior or achieving financial independence, some change in the client's behavior was required. Only clients themselves can make such changes, which are not easy, so they must want to change and be willing to work with the social worker toward that end. For the most part the clients in these projects did not voluntarily seek the social workers' help. There was no indication of what the social workers might have done to motivate them. The latter point is especially important, since "nonvoluntary clients" can often be engaged in a successful change process, but the practitioner must usually make an effort to locate and frame problems they are willing and able to work on. As Perlman commented, "There is no evidence that the caseworkers . . . asked . . . what the client's perception of service might be, and consequently what clarifications and agreements would have to be reached" (1972:194). Or as Wood observed in a latter review of the studies, "Contracts [between the client and the caseworker] do not appear to have existed in the unsuccessful studies" (1978:453). Meanwhile, of course, the evaluators were using measures based on the assumption that caseworkers and clients were attempting to achieve particular kinds of goals.

Finally, questions can be raised about the structure of the experiments. In most there was a strict division of labor between the caseworkers, supervisors, and administrators (program people) on the one hand and researchers on the other. The former conceived and carried out the treatment programs; researchers were cast as evaluators. This structure has the advantage of promoting a more independent, dispassionate evaluation, but it is often bedeviled by problems of communication between program people and researchers, evidence of which can be seen from the mismatch in *Girls at Vocational High* between what the caseworkers may have been trying to do and what the researchers were trying to measure. Also, most of the experiments were one-shot affairs rather than outgrowths of research programs in which controlled experimentation was preceded by one or more pilot tests of the innovative program and outcome instruments. Had adequate pilot testing been done, needed revisions of the intervention program or instruments or both might have been identified.

A few controlled tests of social work intervention conducted during this period did produce some positive findings. For the most part, these experiments avoided at least some of the shortfalls we have reviewed. Two will be briefly discussed for illustrative purposes. In an experiment conducted by Reid and Shyne (1969), a program of planned short-term service limited to eight interviews, had unexpectedly better outcomes than a much longer program of continued service. Although at first glance this might seem to be a negative finding in that the "lesser" service did better, the "lesser," short-term service was actually a different kind of casework with possible therapeutic features lacking in the longer-term treatment program, such as focus on limited goals, the mobilizing effect of time limits, and a more active intervention style on the practitioner's part. Some of these features, in fact, such as limited goals, were in contrast to those of the unsuccessful programs. This was also one of the few experiments in which psychoanalytically oriented casework, which was used in both the short-term and continued service programs, was appropriate to the problems dealt with—largely difficulties in individual and family functioning. (It should be noted that Reid, the co-investigator of that project and the second author of this volume, had nothing to do with the design or implementation of the casework programs!).

Schwartz and Sample (1967, 1972) tested innovative administrative structures for caseworkers in a public welfare setting. An experimental unit in which cases were served by teams of professionally trained and untrained social workers was compared with routine service by untrained workers. Both groups were further split into high and low caseload conditions. Both experimental variables had significant effects on client movement and functioning, with the team approach showing the stronger results. Although Schwartz and Sample found positive results on the kind of global outcome measures that had failed to show differences in other studies, their experiment had the advantage of unitary leadership—the principal investigators were responsible for both the service and research components.

Although a few studies had positive findings, perhaps for the kinds of reasons suggested above and maybe with an added measure of good luck, the field at mid-century was dominated by negative outcome experiments. These failures might be best seen as fledgling efforts to use scientific methods to test the efficacy of social work practice interventions. Like a novice skier who might find it difficult to coordinate skis, poles, arms, and legs, social work found it hard to interweave problems, practice methods, and outcome measures. It was unfortunate that this experimenting was done on such a large scale and became such an embarrassment to the profession.

These efforts failed to contribute the kind of empirically tested intervention knowledge hoped for—to establish the efficacy of certain methods for certain problems. They did, however, raise questions about approaches ill suited to some problems, such as psychoanalytically oriented casework as a means of delinquency prevention. Even though the experiments were flawed, it is reasonable to assume that if the interventions were really powerful, there would have been more positive findings. Thus the experiments were a dramatic illustration of the function of negative findings in knowledge building—to suggest that what has been tested hasn't worked and that something else may be needed.

Moreover, the methodological lessons learned helped set the stage for a new style of field experimentation that emerged in the 1970s and is still the prevailing model. In this style, experiments are directed by investigators with knowledge of both practice and research who have developed and tested intervention approaches. As a result, there have been better connections between intervention goals, methods, and outcome measures. Intervention goals have become less ambitious and methods better suited to the problems tackled (see chapter 4). More emphasis has been placed on research programs with incremental development and testing of both practice approaches and outcome measures (Rothman and Thomas 1994) (chapter 5). As will be discussed further in chapter 7, this new generation of field experiments began to demonstrate the effects of social work intervention and to start producing the kind of positive, empirically based intervention knowledge that the profession had been waiting for for the better part of a century.

Studies of Intervention Processes. Knowledge of intervention effectiveness implies knowledge of intervention processes. To say that a particular social work method proved effective in a study does not provide much useful information unless we are able to state what the method consisted of. For an effectiveness study to be complete, outcome measures need to be combined with intervention measures. More generally, there is a need for empirical descriptions of what social workers actually do as a basis for making assertions about the nature of social work activity.

One of the earliest examples of research on social work intervention was conducted by Mary Richmond as part of her work on *Social Diagnosis* (1917). Richmond actually initiated a series of studies of casework practice. The most elaborate involved "outside sources" (relatives, neighbors, physicians, etc.) used by caseworkers in assessment and treatment planning. Using a carefully drawn purposive sample of 2,800 cases from three cities, Richmond (1917) tabulated sources used. Other analyses were conducted in the 1920s, including those by Robinson (1921) and Myrick (1928). These studies consisted of qualitative analyses of cases to determine the nature of the social workers' activities. They could be seen as elementary forms of content analysis with rough classifications of social work activities. For example, in the Myrick (1928) study, which was conducted by a committee of the Chicago Chapter of the American Association of Social Workers, an effort was made to classify the social workers' methods in a small number of interviews defined as "persuasive." An analysis of one of the interviews produced an ordering of the social worker's techniques, such as "meeting objections through appeals to reason" and "arousing objective attitude toward self" (52). Techniques identified were all case specific. No attempt was made to develop general categories.

A more elaborate content analysis of social work intervention was conducted by Karpf (1931) in his study of family agency records referred to earlier. This was perhaps the first "modern" analysis of social work intervention in that it used a sizeable sample of cases, a general typology of intervention methods, and quantification of findings. For example, Karpf identified "methods used by family caseworkers in their attempts to influence the behavior of clients," including "explanation" (31 percent), "discussion and persuasion" (27 percent), "ordering and threatening" (25 percent), "advice" (10 percent), and "suggestion" (6 percent) (266). Although his categories did not appear to be mutually exclusive, no attempt was made to determine reliability, and statistical analysis was no more sophisticated than the kind of percentage distribution just shown, the study provides one of the few empirically based glimpses of casework in the 1920s.

The increasing shift to psychoanalytically oriented methods that marked direct social work practice during the next three decades raised obstacles to the study of intervention processes. Psychoanalytically oriented casework was complex and difficult to break down into specific categories. Efforts to define and describe casework methods were based more on practice wisdom and reviews of case material than on systematic research (Austin 1948; Family Service Association of America 1953).

A significant breakthrough occurred in the late 1950s and early 1960s in the work of Florence Hollis (1964), who developed and tested a classification scheme that incorporated the kind of psychodynamic methods that caseworkers at the time were using within a more generic system. Her research consisted of a series of studies in which specially recorded process records were coded using her classification method. For the first time in studies of social work intervention, intercoder reliability data were obtained

and used to refine the scheme. Her work sparked a number of derivative investigations using her system as well as other studies using different classification approaches. Soon investigators were making use of tape recordings of casework interviews. (For a review see Fortune 1981). A major contribution of this research was its empirical description of what psychodynamic casework actually looked like in practice. For example, contrary to widely held beliefs, it was found to be far more oriented to clients' current realities than to their unconscious processes or their childhood experiences. Unfortunately, the research on intervention processes conducted during this period had little impact on the outcome studies discussed earlier. As a rule, there was little effort in the latter to collect systematic data on what caseworkers were actually doing.

Infrastructure

The inflow of the kinds of knowledge we have reviewed has depended on the infrastructure social work has used to generate research. We turn now to how that infrastructure evolved and has affected the scientific knowledge produced for the profession.

An Early Marriage and a Quick Divorce. In its earliest beginnings, social work was tied organizationally to the social sciences and thus connected to a nascent scientific infrastructure. Over a hundred and twenty five years ago, when the social sciences and the profession of social work were embryonic, they shared a common belief in social reform and rehabilitation. The group known as the Section on Social Economy that would later become the National Conference on Social Welfare was then part of the American Social Science Association, the precursor of the American Sociological Association. The ASSA promoted the use of the scientific method, attempting to emancipate the study of social problems from the prevailing religious dogma and mysticism. Social work and social science stood briefly together, at least in an organizational sense.

Before long, however, strains began to develop. Early social work practitioners were interested in solving immediate problems, while the social scientists were more concerned with developing research methods and theory (Zimbalist 1977). The practitioners eventually bolted from the American Social Science Association and went their own way, foreshadowing a cleavage between science and social work that exists to this day.

The significance of this very youthful flirtation and parting of ways for the future of social work research is great indeed. The underlying conflict waxes and wanes in succeeding years, but enters to some degree into every relationship between social work and research. The tension between objectivity and commitment, between tentativeness and confident action, between theory building and theory using, which is no doubt necessary and healthy in the practice of most professions, becomes heightened and sharpened when a largely unscientific art is placed against a research framework. By virtue of this early division of fields, social work was largely stripped of scientific skills and manpower and cut off from formal connection with the major sources of theoretical social thought of that era.   (Zimbalist 1977:19)

Social work proceeded to build its own research infrastructure, but the process was slow and uneven. Some research was carried out by social scientists, like Warner, with an interest in scientific charity. Social workers affiliated with settlement houses conducted neighborhood studies. A somewhat ad hoc infrastructure was developed for the social survey movement. Its organizational base was a journal, *Charities and the Commons*, managed by Paul Kellogg, who had been a student at the recently established New York School of Philanthropy. Kellogg secured support from the Russell Sage Foundation to launch the Pittsburgh Survey as primarily a journalistic project. Surveys were staffed by assortments of volunteers and paid "experts," few of whom had formal training in research.

For more general scientific knowledge, social work drew upon research from the social sciences, with occasional studies done by researchers more identified with the profession—e.g., Karpf (1931), Myrick (1928), Theis (1924)—that related to specific concerns of social work. As such studies grew in number, the expression "research in social work" began to be used to refer to studies of social problems that might be targets of social work intervention as well as the intervention methods themselves (Jeter 1937).

Still, the profession had not yet organized a cadre of its own researchers and continued to rely on scientists from other disciplines. For example, in the late 1930s the Community Service Society of New York, the descendant of the New York City COS, decided to develop its own research unit, the Institute for Social Welfare Research. The lead researchers it employed were psychologists and sociologists, although they spent their time on strictly social work issues, such as measuring client movement. The hires reflected not an ignoring of social work talent but a recognition of the lack of social workers with adequate scientific training.

Social workers with research interests and training were becoming more numerous, however. In 1949, they formed the Social Work Research Group (SWRG), the first national organization of social work researchers, which through publications and meetings provided a means of interchange among them (Zimbalist 1977). Its membership, which eventually grew to several hundred, was a mixture of social scientists interested in social work and social workers with research training, often at the master's level. In 1955 SWRG became the Research Section of the newly founded National Association of Social Workers.

One of the projects undertaken by the Research Section was the first major research textbook in the field, *Social Work Research*, edited by Norman Polansky (1960). The title itself was significant, since only in recent years had the term "social work research" begun to replace "research in social work." Zimbalist (1977) views the language change as an important symbolic shift, opening a door for the emergence of an indigenous research enterprise focused on the problems and methods of social work.

In the opening chapter of the Polansky text, Mary MacDonald (1960) provided a succinct definition of that enterprise. "Social work research begins with practical problems, and its objective is to produce knowledge that can be put to use in planning and carrying out social work programs" (3). Her goal, and the general purpose of the other chapters in the volume, appeared to be primarily to define social work research, outline its history and purposes, and create and legitimize an important role for it within social work. The editor wanted to provide a textbook that "practitioners will find understandable and readable" and that teachers of social work and administrators would use as a "review of technical issues" permitting them to make judgments in assessing potential studies (v). The emphasis was on the scientific work that social workers could and should accomplish, not specifically on the methods of borrowing social science knowledge for practice.

Training and Research in the Academy. By the middle of the twentieth century, advanced training for the growing cadre of social work researchers became more available with the development of doctoral programs in schools of social work. Prior to 1940 only three U.S. schools of social work offered doctoral programs; fewer than fifty students in all had graduated from them (Baldi 1971). Beginning in the late 1940s doctoral programs began to mushroom, and at major schools, such as the New York School of Social Work. By the mid-1960s nineteen programs were in place. However, the number of graduates remained small—for example, from 1950 through 1968, the programs in total produced fewer than thirty graduates a year (Baldi 1971).

Another element in the evolving infrastructure of social work research was sponsorship of research activity. Until the mid-twentieth century, most social work researchers were employees of social agencies. Some of the more important studies were undertaken by special research units in large or national organizations, such as the Institute for Welfare Research of the Community Service Society and research departments of the Family Service Association of America and the Child Welfare League. As doctoral programs began to build a professoriat with research training and interests, research leadership began to shift to academic settings. Social agencies continued, of course, to be sites for research involving client populations, but increasingly such research was directed by academic investigators, many of whom combined a practice background with research training, which gave rise to the advances in field experiments discussed earlier. The increasing control of scientific knowledge building by the academic arm of the profession raised the issue of whether academic researchers were in sufficient touch with the realities of practice to provide the best kind of leadership.

The growth of a professoriat with doctoral training also meant a growth in research activity. Whether motivated by their scholarly interests or responding to university expectations that they be "productive," the new academic researchers began to increase considerably the volume of social work studies. The projects were also of higher quality, by and large, than earlier agency-based efforts by researchers without advanced training and without the benefit of university resources.

Finally funding for research became more diversified and more available. In its early decades social work relied largely on social agency budgets and private foundations to support its research efforts. With the creation and expansion of federal research programs, such as the National Institute of Mental Health, that began in the 1950s, government sources of support became dominant, with other sources serving a supplemental role. The influx of government dollars, usually in much greater quantity than could be obtained from other sources, stimulated a good deal of social work research activity during the middle and later decades of the twentieth century. However, there were strings attached. Increasingly the funding agencies themselves began to determine the priorities for the kinds of knowledge to be developed, which did not always appear to fit the needs of the profession.

Summary and Conclusion

From its beginnings social work attempted to make use of science as a base for its practice. The hope that science could advance knowledge and methods in social work as it had in medicine and other disciplines could provide a claim to professional status were among the driving forces. The adoption of the scientific method as a model for practice became the leitmotif for late nineteenth-century charity work, was advanced considerably a few decades later in the monumental contributions of Mary Richmond, and remained as an ideal during the decades of psychoanalytic dominance. Although perhaps seen better as rational problem solving than science and as more of an aspiration than a reality, the model nevertheless set a standard for best practice and prepared the way for the more elaborate scientific practice models that would later evolve.

The earliest forms of scientifically based knowledge were case specific, generated by systematic gathering of data about individuals or families, as in the application of the protocols of *Social Diagnosis*, or about communities, as in the social survey. More general, and ultimately more powerful, forms of scientific knowledge developed much more slowly, and not in a form that social workers could effectively use; knowledge from psychoanalysis was readily usable, but its scientific warrants were questionable. Efforts to develop intervention knowledge through evaluations of casework programs yielded meager returns. Systematic knowledge about the nature of social work practice and methods of studying it, essential to building intervention knowledge, did not emerge until the middle of the twentieth century, and not until the final decades of that century had sufficient general information relating to assessment and intervention accumulated to make a difference in social work practice.

From the late nineteenth to the late twentieth century, the development of tools for building scientific knowledge was probably of greater significance than what they produced. Progression in scientific method was a pervasive theme during this period. The capacity of science for self-correction was evident in a wide range of methodologies—the social survey, explanatory investigations of psychosocial problems, experimental tests of social work interventions, and studies of social work processes. These advances eventually began to make scientific knowledge a force in social work practice.

To build scientific knowledge, a profession needs a research infrastructure that includes a cadre of researchers, a domain for their activities, organizational structures to enable them to communicate with one another, a means of providing them with specialized training, auspices to enable them to conduct research, and sources of funds for their efforts. There was little development of this kind of infrastructure in social work until the middle of the twentieth century. At that point a number of factors coalesced, including the establishment of organizational structures, the identification of a domain (social work research), the rapid development of doctoral programs, the increase of government funding, and the shift of the research enterprise from the agency to the academy. The emergence of this infrastructure was more abrupt than incremental and relatively recent. It led to advances in the kind of research conducted as well as to a large increase in the quantity of studies produced. But it also raised issues concerning relations between the profession's growing research arm and the practice community.

CHAPTER THREE

Client Problems as Organizing Foci for Knowledge*

One natural link between practice and science is a shared propensity among social work practitioners and social scientists to focus on the problems that people experience as the starting point for inquiry. To be sure, social workers are more interested in helping clients to cope, whereas social researchers want to better understand the causes and correlates of client problems. Nevertheless, problem assessment is the focal point for organizing knowledge about people in trouble. This has led naturally to the delineation of different types of client problems.

Researchers, more than practitioners, have stressed the need for formal classifications of client problems. Researchers' interest in typologies of problems is easy to understand. Fundamental to science is making meaningful distinctions among phenomena, documenting the significance of these differences, grouping things into classes, and developing methods of reliably earmarking them as belonging to one or another group. Scientists then try to develop knowledge about these distinct entities. Classifications are initial structures for building and accessing scientific knowledge.

Tracing how social work has attempted to use science to inform and shape practice requires considering how both practitioners and researchers

*Parts of this chapter were presented at a conference on developing practice guidelines held at the George Warren Brown School of Social Work, Washington University in St. Louis, May 3–5, 2000 and will be published in Proctor and Rosen, in press.

have approached client problems. This chapter will review how this mutual interest in client assessment has evolved. First, we will revisit how problem diagnosis and classification have been discussed traditionally in social work. Second, we will describe several recent, prominent, formal problem classification systems (namely, the *Diagnostic and Statistical Manual of Mental Disorders* and the Person-in-Environment system) and how they appear to depart from the earlier tradition. Finally, we will evaluate a contemporary attempt to de-emphasize problem classification as the organizing focus of intervention knowledge in favor of treatment objectives.

Assessment and Purpose in Social Work

Mary Richmond saw assessment not as a decision about the placement of an individual into a diagnostic category but as a process of gathering information systematically, making inferences carefully, and developing a plan of intervention. In her book, *Social Diagnosis* (1917), assessment of clients is identified as a central task of social casework. Richmond argues, in fact, that the methods of diagnosis should constitute part of what makes social work unique, "a part of the ground which all social case workers could occupy in common" (5). Her early formulation, cumbersome as it may appear today, emphasized the need to develop a comprehensive understanding of the client's problem as a basis for treatment (Kirk, Siporin, and Kutchins 1989; Mattaini and Kirk 1991). Her perspective can be found throughout the social casework literature of the twentieth century.

One representative example of the thinking about assessment from mid-century casework literature is Helen Perlman's 1957 book, *Social Casework: A Problem-Solving Process*, in which she recognizes the variety of problems that draw the attention of social workers.

> There is probably no problem in human living that has not been brought to social workers in social agencies. Problems of hunger for food and of hunger for love, of seeking shelter and of wanting to run away, of getting married and of staying married, of wanting a child and of wanting to get rid of a child, of needing money and of wasting money, of not wanting to live and of not wanting to die, of making enemies and of needing friends, of wanting and of not wanting medication, of loving and of being unloved, of hating and of being hated, of being unable to get a job and of being unable to hold a job, of feeling afraid, of feeling useless—

all these, and the many other problems of physical and emotional survival as a human being, come to the door of the social agency.

(Perlman 1957:27)

In this passage, Perlman is not proposing a problem classification system (although her description might serve as well as most); she is emphasizing the diversity of clients and their struggles and misfortunes. In fact, in her chapters "The Problem" and "Diagnosis: The Thinking in Problem-Solving," Perlman barely mentions classification, although she has much to say about client problems.

She makes clear that there is almost never just one clearly defined problem. First, she says, there is *the problem that the client wants help with*, the one that serves as the impetus to seek a social worker. Second, there is *the problem identified by the caseworker* based on professional knowledge and judgment. This is not one but a multilayered series of problems. Perlman uses a case example (31–32) of an overwhelmed mother whose baby was hospitalized in a diabetic coma to describe how the caseworker might identify the "basic problem" (possibly the mother's neurotic character disorder), the "causal problem" (the mother's parental relationships), the "precipitating problem" (the baby's illness), the "pressing problem" (the baby's imminent release from the hospital) and the "problem-to-be-solved" (the mother's insecurity about managing the sick child). And, finally, there is the *purpose(s) of the social agency*, which directs the attention of the caseworker employed there.

In addition, Perlman reminds us that for every client problem there is both an "objective" and a "subjective" significance (35), the way the problem looks to an observer and the way it feels to the client. Furthermore, whatever the nature of the problem that the client brings to the agency, "it is always accompanied, and often complicated, by the problem of being a client" (37), of help seeking and help taking. She concludes that "The problem brought by the person to the agency, then, is likely to be complex, ramified, and changing even as it is held to analysis" (39). The elegance of her dissection of the client's "problem" should have served as a warning to those attempting (unsuccessfully) decades later to develop formal problem classification systems.

In her chapter on diagnosis as part of the problem-solving process, Perlman has an equally insightful view. The "*diagnostic process*" involves the caseworker marshaling the facts of person, problem, and place; analyzing and organizing them; reflecting upon them; and making judgments of their meaning to determine what to do and how to do it. The conclusion of this process, stating what the trouble seems to be, how it is related to the client's goals, and what the agency, caseworker, and client can do about it constitutes the *"diagnostic product"* (164; emphasis added). "Diagnosis," she states, "if it is to be anything more than an intellectual exercise, must result in a 'design for action'" (164). In her view, it is tied to intervention:

> We seek to clarify the nature and configurations of the material (person-problem-place-process) that we are attempting to influence . . . we try to organize our half-felt, half-thought impressions into some conclusions, temporary though they may be, that will give direction for what to expect and do next. (165)

She describes diagnosis as a process of organizing "intuitions, hunches, insights, and half-formed ideas," putting them "together into some pattern that seems to make sense . . . in explaining the nature of what we are dealing with and relating it to what should and can be done" (166). "The content of casework diagnosis is focused, weighted, and bounded by the *purpose* and *means* of the client and the agency" (169; italics in original). She goes on to describe several different "kinds of understanding" (170) that she labels dynamic diagnosis, psychosocial diagnosis, clinical diagnosis, and etiological diagnosis. Only in regard to clinical diagnosis does she mention classification, as an attempt to categorize a person by the nature of the sickness, and she relegates it to the province of psychiatry, not social casework. The purpose of diagnosis (understood as encompassing all four types) in social casework, she concludes, "is to give boundary, relevance, and direction to the caseworker's helpful intents and skills" (179).

Other early texts also emphasize that social work assessment strives for an understanding of the client's problem as a guide for treatment and that classification per se is not particularly important (Hamilton 1951:213–236). The purpose of the diagnostic formulation is to provide a foundation for treatment. Similarly, Florence Hollis's *Casework: A Psychosocial Therapy* (1964), in the chapter, "Assessment and Diagnostic Understanding," takes a broad view, minimizing the need for problem classification in favor of a comprehensive understanding of the client and the situation. The book notes that there is medical classification of physical diseases and clinical diagnosis in psychiatry but has little to say about problem classification beyond indicating that it is important to identify whether the client's problem involves "family dysfunction, marital conflict, parent-child problems, unwanted pregnancy, delinquency, substance abuse, unemployment, old age, premature births, incest survival, and situations involving children of alcoholics and terminal illness" (259). Classification in any formal sense is not pursued.

By the early 1970s, however, diagnosis and classification were given more extended, if skeptical, attention by Scott Briar and Henry Miller in *Problems and Issues in Social Casework* (1971). Reviewing the casework literature, they observe that classification should allow for the development of better treatments but note that classification schemes in psychiatry and elsewhere have not been particularly reliable or valid. They suggest, furthermore, that the usefulness of diagnosis is limited by the specificity of the treatment available.

> If there were only one method of intervention to be applied in all cases, diagnosis would be superfluous, since discrimination among cases would be unnecessary—all cases would be treated with the same method anyway. Thus, Carl Rogers' client-centered therapy required no diagnostic activity, since the same therapeutic approach was used with all clients.   (145)

The Briar and Miller book, appearing during casework's effectiveness crisis and authored by two prominent scholars at Berkeley, served as a critique of theory and practice and a plea for a more behaviorally oriented casework grounded more solidly on empirical data. The authors appear to favor an inductive approach to classifying clients according to their responsiveness to specific treatments (147–151). First, they suggest, we need to discover who improves with a specific treatment; then we must learn what else those clients might have in common. From this inductive effort, a typology of clients could be developed and matched with treatments most likely to help them. This suggestion was generally ignored by the field.

Problem Classification and Purpose

No coherent argument can be made for or against classification systems in general. The usefulness of any system depends on whether its intended purpose is advanced. Classification is fundamental to the way in which humans organize information about the world and their experiences. To classify is simply to assign persons or things to a group by reason of presumed common attributes, characteristics, qualities, or traits. In biology, classification involves a system of categories distinguished by features like structure or origin. Physical, mental, and social human maladies have been labeled and grouped for centuries by etiology, pathological processes, or observable symptoms. Diagnosis is commonly a form of classification according to the nature and circumstances of a diseased condition.[1]

Classification of client problems or human efforts always produces simplification, because it forces attention to selected characteristics as the primary basis for grouping, ignoring more information and more diversity than it ever captures. Thus, it is a decision to ignore some information on the hunch that the characteristic chosen as the basis for assignment to groups will fulfill some central purpose. For example, grouping people as male or female highlights some physiological characteristics but ignores thousands of other human traits. Whether it is useful depends on its purpose. For example, classification by gender would be useless for identifying intellectual ability, but not for identifying those most likely to get prostate or breast cancer.

Classification in social work always has this issue of purpose as a backdrop. There are four common purposes of classifying human maladies. The first is *administration*, to sort the needy from the not needy, the eligible from the ineligible, the competent from the incompetent. Many of the public debates about social welfare policy revolve around these distinctions. The second common purpose is *triage*, to give some of the eligible and in need priority for the allocation of scarce treatment services. This involves debates about who is most needy or what the cost-benefit ratio may be for different allocations of resources. The third common purpose is *knowledge development*, particularly in the quest to understand etiology. Those engaged in basic research pursue the causes of maladies and need to identify the afflicted and the nonafflicted in order to scrutinize their genes, psyches, habits, and social circumstances for correlates of affliction that may lead to clues about cures or prevention. Finally, the fourth common purpose of classification is to more effectively *help particular clients*.

With regard to this latter purpose, there are currently two prominent classification schemes: the system created in psychiatry, which has pro-

---

[1] "Diagnosis" can have other meanings, including "the process of determining the nature of a condition; the description that classifies a phenomenon; the identification of a cause or nature of a problem; and the answer or solution to a problem" (*Random House Unabridged Dictionary*, 2nd ed.).

duced various editions of the *Diagnostic and Statistical Manual of Mental Disorders*, and the social work system called Person-in-Environment (PIE). (In both psychiatry and social work, there have been many less successful schemes advanced over the years.) We will examine each of these briefly, since they exemplify approaches to organizing assessment knowledge. We will see how they diverge from some social work traditions and how they warn about common pitfalls of formal classification systems.

DSM

Nowhere in the general purview of social work is there a more developed and institutionalized system of problem classification than the American Psychiatric Association's *Diagnostic and Statistical Manual of Mental Disorders (DSM)*. *DSM* is, of course, a classification of more than 300 mental disorders that has appeared in its modern versions in 1980, 1987, and 1994 (APA 1980, 1987, 1994). Extensive reviews of *DSM* can be found elsewhere (e.g., Kirk and Kutchins 1992; Kutchins and Kirk 1997; special issues of the *Journal of Abnormal Behavior*, August 1991 and August 1999). Here, we merely need note that *DSM* is the best of its kind, a product of decades of labor by hundreds of dedicated psychiatric researchers and professionals who had to contend with many conflicting professional interests.

The classification of mental disorders has not always been so labor intensive. Although descriptions of madness and its subtypes have been around since the ancient Greeks (Zilborg 1941; Alexander and Selesnick 1966), until the last half of the twentieth century, a handful of unofficial, broad categories appeared sufficient for the task. The recent evolution of psychiatric classification is the linguistic legacy of nineteenth-century epidemiology, which pursued the causes of infectious diseases by plotting morbidity and death among various populations (Mirowsky and Ross 1989a, 1989b). These methods presupposed that people could be sorted into two groups: those with the disease and those without. Psychiatry, mimicking the approach of medicine, adopted this basic concept of "caseness" and developed classifications into which cases of psychiatric disorders could be placed.

The earliest classification systems of mental disorders in the United States were developed by the federal government for the U.S. Census, which played a predominant role in psychiatric nosology for almost a century (Grob 1991). By the 1880 census, there were seven official categories of mental disease: mania, melancholia, monomania, paresis, dementia, dipsomania, and epilepsy. In 1904 and 1910, two special surveys were conducted enumerating the institutionalized insane. The one in 1904 was particularly concerned with race and ethnicity, reflecting the growing fear of large-scale immigration of presumably inferior groups, which might somehow be documented by statistical studies of patients in asylums (Grob 1991).

In subsequent years, census officials became more interested in the need for a standard nosology and asked the American Medico-Psychological Association, the forerunner of the American Psychiatric Association, to appoint a committee on nomenclature to facilitate the collection of data. Psychiatrists who were involving themselves in a broader array of community problems beyond the administration of mental hospitals, adopting a broader vision of their social mission, began to see how social statistics might be used to guide mental health planning. By 1918 the first standardized psychiatric nosology, the *Statistical Manual for the Use of Institutions for the Insane*, was produced. It offered twenty-two principal categories that had a decidedly somatic or biological orientation. This was congruent with the fact that most psychiatrists practiced in mental hospitals and many patients, perhaps a majority, had severe physical as well as mental problems. The somatic nosology reflected the nature of psychiatric care (Grob 1991). The manual was adopted by the census and used to survey mental institutions annually, a tradition that was continued after World War II by the newly established National Institute of Mental Health.

The major opponent of the new nosology was Adolf Meyer, a prominent figure in American psychiatry, who said that

> statistics will be most valuable if they do not attempt to solve all the problems of administration and psychiatry and sociology under one confused effort of a one-word diagnosis marking the individual. . . . The statistics published annually as they are now are a dead loss to the States that pay for them, and an annual ceremony misdirecting the interests of the staff.   (quoted in Grob 1991:426)

Despite such criticism, the new manual went through ten editions between 1918 and 1942, retaining its somatic orientation. In 1935, it was incorporated in the first edition of the American Medical Association's *Standard Classified Nomenclature of Disease* (Spitzer and Williams 1983).

These advances in nosology, however, were of only marginal significance to psychiatrists and their patients. The categories were broad, and psychiatric treatment at the time was nonspecific. The struggles to develop a systematic nomenclature, from the earliest decades of the nineteenth century, were motivated by administrative and governmental needs, not by demands from practitioners (Grob 1991). This is a pattern that has persisted.

The experience of psychiatrists during World War II was responsible for the first major change in psychiatric nosology. It was embodied in the *Diagnostic and Statistical Manual for Mental Disorders*, now commonly referred to as *DSM–I* (APA 1952). Published in 1952 by the American Psychiatric Association, it reflected major political and theoretical shifts in American psychiatry. The somatic tradition gave way to psychodynamic and psychoanalytic perspectives, ascendant in the profession by the middle of the twentieth century. These new viewpoints emphasized the role of environment in mental disorders and the variety of less severe forms of disturbance that could benefit from the attention of psychiatric professionals. Following the war, there was a dramatic increase in the number of psychiatrists and a radical shift in the settings where they practiced, away from mental hospitals to community clinics and private practice (Grob 1991). Clinicians increasingly worked with noninstitutionalized populations and those suffering from less severe disorders, such as neuroses and personality disorders, rather than psychosis. A more "modern" nosology was needed. *DSM–I* reflected this transformation of psychiatry and the ascendancy of new leadership in the profession. It marked a triumph of psychodynamic perspectives over the older, organic, institutionally based nosology (Grob 1991).

DSM–II

DSM–II (APA 1968) was a small, spiral-bound notebook that clinicians could purchase for $3.50. It expanded the number of disease categories and continued the psychodynamic traditions of *DSM–I*. By contemporary standards, making *DSM–II* was a relatively private and simple process, more like changing rules and regulations within one organization than negotiating treaties among many rival factions with very different objectives. The presentation of *DSM–II* to the mental health community was couched in language that suggested nothing substantive had been changed. Yet new categories of disorder were added, the nomenclature was organized in a different way, the recording of multiple psychiatric diagnoses and associated physical conditions was explicitly encouraged, qualifying phrases were changed, and numerous definitions of disorders were revised (Spitzer and Wilson 1968). In fewer than 40 pages, *DSM–II* offered the definitions for 182 specific disorders. The one-paragraph description of schizophrenia—the pivotal disorder from which modern psychiatric classifications have evolved—illustrates the brevity of these definitions. Even the more extended discussion of subtypes of schizophrenia was less than a page long. There was no attempt to justify the many changes on the basis of scientific evidence; like earlier nosologies, *DSM–II* was intended primarily to reflect, not to change, the current practice of psychiatry. Its publication drew no public attention or concern.

Despite its noncontroversial origins, *DSM–II* quickly became a convenient scapegoat for those who wanted to criticize American psychiatry. The second edition was vague, inconsistent, theoretically clumsy (despite earlier claims that it eschewed theory), and empirically weak in terms of both the validity and the reliability of its diagnostic categories. Nevertheless, for psychodynamic therapists who predominated in American psychiatry after World War II, *DSM–II* served as a comforting and familiar small manual. Many clinicians appreciated its simple administrative uses; few viewed it as a treatise on psychiatric philosophy or treatment (Grob 1990; Wilson 1993). As a minor desk reference, *DSM–II* was neither prominent nor controversial among practicing clinicians. For psychiatric researchers, however, it was nearly useless as a scientific guidebook. This problem led to the extensive revision of the manual that has been characterized as transforming American psychiatry in this century (Wilson 1993).

DSM–III

All the editions of *DSM* since 1980 have been explicitly designed to be "descriptive" and "atheoretical." They provide the observable signs and symptoms that a clinician should find in order to determine whether a client has a particular mental disorder. These recent editions of *DSM* make few explicit assumptions about etiology (most of which is in dispute anyway), in contrast to *DSM–II*, which had psychoanalytic concepts and assumptions embedded within its text. In defining categories of disorder, *DSM–III* uses methods borrowed directly from psychiatric researchers— each disorder is operationalized by a list of necessary and sufficient observable "diagnostic criteria." Their purpose is twofold: to increase the specificity and validity of the diagnostic system by ensuring that only those with a disorder are properly labeled, and to eliminate clinical discretion (and error) as much as possible to improve reliability.

In making *DSM* atheoretical with respect to etiology, the developers also wanted to make it neutral in regard to recommended treatment. The system's architects recognized the sharp conflicts in treatment philosophy among psychiatrists and other mental health professionals and wanted to avoid aggravating them (see the fine analysis by Luhrmann 2000). The political process of gaining approval of the various editions of *DSM* was complex enough, with battles waged over such issues as the proposals to eliminate the term "neurosis," to reinsert a diagnosis for homosexuality, and to create diagnoses for masochistic personality disorder, tobacco and caffeine dependence, and other conditions. Thus, *DSM* explicitly rejected the casework premise that diagnosis was an incipient form of treatment planning.

DSM's success as an official classification system is due to its usefulness for very different practical purposes. Among clinicians, it is primarily a codebook used to qualify for insurance reimbursement for treatment (Kutchins and Kirk 1988; Kirk and Kutchins 1988). For managed care companies, its widespread acceptance facilitates decision-making and administrative work. Among researchers, *DSM* is a handy (and now required) screening tool for selecting people to participate in studies of etiology and treatment effectiveness. Among educators, it provides a rubric for teaching students by organizing information about different kinds of personal difficulties. And finally, for all mental health professions, it provides a sense of rigor and scientific respectability in psychiatric diagnosis, which had for decades been the whipping boy of critics. For our purposes, it is worth noting that *DSM*'s singular focus on defining internal psychopathology is radically different from social work's general historical concern with family, kinship, culture, and institutional context in assessment and treatment (Kirk, Siporin, and Kutchins 1989). Furthermore, *DSM* is not a set of prescriptions for action.

PERSON-IN-ENVIRONMENT (*PIE*)

The quest for scientific and professional respectability was the key factor in the development of a problem classification system by social workers in the mid-1980s, with some support from the National Association of Social Workers. To this day, efforts continue to promote the Person-in-Environment system (PIE). Unlike *DSM–III*, PIE was not a revision or replacement of an earlier system; it was a new endeavor. That is not to say that there were no prior attempts to classify the problems of clients. Many specific agencies and individual scholars and researchers attempted to categorize and describe client problems (see Karls and Wandrei 1994:6–11 for references to some of these). But these classification systems were often limited to specific administrative or research purposes or were a tangential part of a larger scholarly endeavor, and never took hold much beyond the developers' immediate purposes. Both the American Orthopsychiatric Association and the American Psychological Association have flirted with the possibility of developing counterparts to *DSM*, without success.

Against this background, James Karls and Karin Wandrei (1994) worked for many years to develop and promote the Person-in-Environment (PIE) system of classification. They introduce the system by claiming that it was an "important step in the development of the profession" (xvii) that would lead to "a universal classification system" (xvii).

> The social work profession has long struggled to establish its identity and to assert its independence and uniqueness among the human services professions. For lack of a common system of communication, social work has had to rely on the systems of other professions to describe its clientele. For example, it uses the language of psychiatry in working with persons who are mentally ill or emotionally troubled, the language of general medicine when working with persons who are physically ill, and the language of law when dealing with penal and civil code violators and their families. Social work must acquire its own language and its own nomenclature to describe its unique area of service. In so doing, social work will establish a clearer definition of its areas of expertise and will also establish itself as a major profession. (Karls and Wandrei 1994:4)

They admit that they are "unabashedly ambitious" for social work, which should be recognized

> as *the* profession that can best help with the social functioning problems that befall us all. We are convinced that, for this to happen, social work must acquire and use its own language to communicate the work it does and the role it plays in eliminating or alleviating problems in the human condition. We are advocating that PIE be that language.
>
> (xvii; emphasis in original)

The incredible success of *DSM* had stimulated Karl and Wandrei's search for a new language that would capture social functioning rather than internal psychopathology. PIE was an ambitious reaction. But what the new system was attempting to accomplish and how was unclear. These confusions are apparent in the opening paragraph of chapter 1:

The person-in-environment (PIE) system describes, classifies, and codes the social functioning problems of adult clients of social workers. Using the organizing construct of person-in-environment, PIE creates uniform statements of social role, environmental, mental, and physical health problems; and client strengths. The system seeks to balance problems and strengths; it delineates problems pertinent to both the person and the environment and qualifies them according to their duration, severity, and the client's ability to solve or cope with them. PIE is intended for use in all current fields of social work practice and by practitioners of varying theoretical positions. It is not a diagnostic system because it does not offer a cause-and-effect relationship for the problems identified. Instead, it is a tool for collecting and ordering relevant information that can produce a comprehensive assessment of a client's problems in social functioning, which in turn allows for the application of interventions from varying theoretical positions that might relieve or solve the problems presented.   (3)

How these many concepts and purposes were to be conceptually linked to form a new system of communication is not explained in this exuberant statement. Constructing PIE involved borrowing disparate concepts from social work and the social sciences, such as social functioning, social role, client strengths and coping ability, and community environment. Unfortunately, this introduced confusions of purpose. For example, in the quote above, the developers declare that PIE is "not a diagnostic system because it does not offer a cause-and-effect relationship for the problems identified" (3). This is a curious disclaimer, because neither *DSM* nor medical classifications claim that their categories are based on etiology—if that is what "cause-and-effect" means in this context. Diagnostic systems don't require cause-and-effect explanations, although they are obviously strengthened if they include them.

Like *DSM*, PIE has many categories arrayed around distinct "axes." There are types and subtypes, ratings of severity, and code numbers. The PIE system at first appears to be *DSM*'s multiaxial system turned upside down, in that it elaborates the interpersonal and social aspects of a client's life far more than do the simple rating scales reserved for these matters on axes IV and V of *DSM–III*. The 1988 PIE training manual gives an example of the application of the system. The client was a twenty-five-year-old hospitalized patient with AIDS. Within three weeks, he had lost his job, and consequently his health insurance, and was no longer able to support himself or

TABLE 3.1 Basic Structure of PIE[*]

Factor I: Social Functioning Problems

A. Social role in which each problem is identified (four categories)
B. Type of problem in social role (nine types)
C. Severity of problem (6-point indicator)
D. Duration of problem (6-point indicator)
E. Ability of client to cope with problem

Factor II: Environmental Problems

A. Social system where each problem is identified (six systems)
B. Specific type of problem within each social system (number varies for each social system)
C. Severity of problem (6-point indicator)
D. Duration of problem

Factor III: Mental Health Problems

A. Clinical syndromes (Axis I of *DSM—IV*)
B. Personality and developmental disorders (Axis II of *DSM—IV*)

Factor IV: Physical Health Problems

A. Diseases diagnosed by a physician (Axis III of *DSM—IV, ICD–9*)
B. Other health problems reported by client and others

[*]Taken from Karl and Wandrei 1994:24.

to afford medical care. He was forced to reveal to his family that he was gay, which caused a great deal of conflict. In addition, there was panic among the hospital employees, who were afraid of becoming infected. He was severely depressed.

This seems like a case tailor-made for Helen Perlman's discussion of types and layers of problems. PIE renders it in this manner:

Factor I. 3170.442 Worker Role—Paid Economy Problem, Separation/ Loss Type: High Severity, 2–4 weeks duration, Adequate Coping Skills (Presenting Problem)

1320.442 Child Role Problem/Ambivalence Problem: 2–4 weeks duration, Adequate Coping Skills

Factor II. 5405.44 Economic/Basic Needs System Problem: Economic Resources—Other (insufficient economic resources in the community to support self and provide needed services): High Severity, 2–4 weeks duration (Presenting Problem)

8412.24 Health, Safety, and Social Service System Problem: Discrimination on the basis of health status, Low Severity, 2–4 weeks duration

Factor III. Axis I. 296.33 Major Depression, Single Episode. Axis II V71.09 No diagnosis on Axis II.

Factor IV. AIDS (By Dr. Y)

(Karls and Wandrei 1988)

On the basis of this example, it is tempting to dismiss PIE as an overelaborate mystification of human tragedy. It seems to fail in its stated purpose of attempting to provide a clearer language for communication among practitioners and to devise a shorthand for problem description. The formal diagnosis is longer than the case description itself. Like *DSM*, PIE gets bludgeoned by its own parameters, pushing practitioners into thinking that placing clients in complicated categories is the clinical objective, assigning obscure code numbers is the scientific goal, or filling out multiaxial administrative forms constitutes the stepping stone to professional status.

To be fair, PIE undertook much greater tasks than *DSM*. Unlike *DSM*, which explicitly encourages clinicians to use existing category labels consistently, PIE created and defined categories where none existed and where there was relatively little empirical foundation for them. Furthermore, describing internal psychopathology is simpler than describing social functioning and environmental problems, which requires covering much more varied conceptual territory. Unlike *DSM*, PIE was not trying to restrict clinical decision making but to reshape social work's entire jargon, even though there was no institutional structure or existing sanction to support such an effort. Where *DSM* uses diagnostic criteria to guide inferences about the presence of internal psychopathology, PIE's categories are descriptions of the problems using a "new language." *DSM* tries to eschew references to intervention, but PIE encourages them, although it does not systematically link them to the problems (see a case example in Karl and Wandrei 1994:40). Unlike *DSM*, PIE is designed for practitioners, not researchers; it comes out of practice experience rather than laboratories and is therefore less grounded in the research literature or traditions. Finally, PIE is a small-scale endeavor, championed by a small group of supporters on a shoestring budget, whereas *DSM* is a massive effort with access to many scientific and political resources.

Whatever the eventual fate of PIE, to date it has not yet had any appreciable influence on research endeavors in social work and thus has not served to codify assessment knowledge. Recent editions of *DSM* emerged from the research community but attempted to be sensitive to some concerns of practitioners; PIE has had little connection with the research community, relying on a few active promoters. What we learn from PIE is that small-scale efforts to reshape or restructure conventional social work practice or knowledge are likely to be fraught with difficulties.

Organizing Knowledge for Intervention

Almost all of the early casework theorists viewed assessment as an attempt to understand the nature, meaning, and evolution of the client's problem and to develop a plan of action on the basis of that understanding. As classification systems and guides to organizing knowledge, both *DSM* and PIE have distanced themselves from (if not completely denied any interest in) links to etiology, meaning, and intervention.

This neglect of intervention in assessment of clients would strike many of the early casework theorists as lopsided. As Perlman indicated, clients arrive at social agencies with problems of many kinds. The social worker attempts to understand the nature of the presenting problem and then develop with the client some treatment goal (objective, target) and select some method (intervention) to achieve it. This is all part of the therapeutic relationship. The treatment goal or target may be to alter the contributing causes, to alter directly the client's presenting problem, or to alter some secondary undesirable effect of the client's problem (see figure 3.1). For example, interventions may attempt to remove, modify, or ameliorate one of the contributing causes of the client's presenting problem. In the case of a child who is being maltreated, the worker may help the parents improve their parenting skills so that they will not neglect or abuse the child (Target 1). Or the worker could focus efforts on removing, modifying, or ameliorating the presenting problem itself rather than a contributing cause. For example, the abused child could be removed from the home (Target 2). Or the worker might move first to ameliorate the effects of the problem by arranging medical and psychiatric care for the traumatized abused child (Target 3).[2]

---

[2]It should be noted that in all these cases, the intervention is preceded by some assessment of the problem, regardless of which aspects of that assessment guide the choice of treatment objective.

FIGURE 3.1

Contributing Causes →	Client Problems →	Secondary Undesirable Effects
(multiple)	(multiple)	(multiple)
∧	∧	∧
Intervention	Intervention	Intervention
Target 1	Target 2	Target 3
Change Parental Behavior	Find Safe Placement	Treat Child's Medical Problems

Because of the attention given to assessment and diagnostic systems as organizing schemas for practice knowledge—particularly in the development of practice guidelines (see chapter 8)—and the recent relative lack of emphasis on plans of action, several scholars have advanced a proposal to organize practice knowledge according to treatment objective. Currently, practice knowledge and intervention guidelines in medicine and psychiatry are organized by diagnosis or problem, e.g., schizophrenia, depression and so on. Enola Proctor and Aaron Rosen of Washington University (Proctor and Rosen in press) have recently proposed that knowledge and guidelines be organized on the basis of "what the interventions are supposed to achieve"—the targets or goals of intervention—rather than on the basis of the client's problem. They reason that since intervention often does not attempt either to reverse or remove the problem, codifying knowledge in terms of the problem is not very useful.

From the practitioner's perspective, the first requirement for use of guidelines is to have an efficient means to enter the compilation of knowledge statements and connect with the knowledge that is relevant for the task at hand. To be maximally useful, therefore, guidelines for intervention should be organized around the targets toward which interventions are directed . . . all professional interventions aim to achieve some desired outcomes. . . . Hence, outcome-based concepts . . . are the appropriate concepts for a nosology guiding access to practice guidelines. (3)

For example, when responding to a client seeking help because of spousal abuse (the problem), the worker may well choose to place the client in a shelter (the outcome) rather than change the spouse's abusive behavior. . . . When an outcome is not a reversal of the problem, we may need to affect a condition that is substantively different from that of the problem, and the intervention capable to achieve the desired outcome is likely different than the intervention that would resolve the problem. (6–7)

On the surface, at least, the proposal to organize intervention knowledge by target has considerable appeal. It appears to provide a practical handle for social workers to grasp in deciding how to intervene most effectively. For example, a worker who needed to place a battered woman in a shelter might examine a practice guideline organized under an index term like "finding shelter" for a useful summary of knowledge about what to do to achieve this goal. Target-based guidelines would be forward-looking (how to get from here to where you want to go) rather than backward-looking (how the client got here). The proposal is, in part, an attempt to redress the recent neglect of a plan of action.

Would it really be better to organize social work intervention knowledge around goals to be achieved rather than problems that have already occurred?[3] Proctor and Rosen recognize that problem analysis can't be totally abandoned; "diagnosis and problems do play a legitimate role in treatment considerations. The question is, What role should they play?" (personal communication 12/2/99). They argue, however, that practitioners already often choose interventions based on the goals they want to achieve rather than on the problem classification; therefore, guidelines should be organized around the clinical targets.

For the sake of discussion, we will assume that intervention knowledge could be organized around intervention objectives, that practice guidelines could be formulated on the basis of this knowledge, and that practitioners would be motivated and able to access these guidelines. Under those assumptions, what would be the role of diagnosis or problem classification?

It is possible that goals can be achieved, such as parents given parenting skill training, children placed in safe environments, and youths provided with medical attention, without great effort expended to understand or identify a focal problem. Let us consider this by using the extreme hypo-

---

[3]We need to be mindful of an important distinction between the tasks of the practitioner deciding how to intervene with a particular client and those of the scholar attempting to organize knowledge. The role of problem classification in each case needs to be addressed separately.

thetical test of trying to select targets and interventions without any problem identification. Could practitioners dispense with diagnosis, problem classification, or assessment altogether?

SELECTING TARGETS WITHOUT IDENTIFYING PROBLEMS

Let's take the case of the battered woman who needs shelter, in which the target (the placement) does not involve altering the spouse's abusive behavior (the immediate cause of the woman's problem). How was shelter placement selected as the target? In this brief example, of course, we really don't know, but this particular outcome is not the only or obvious target when a woman has been abused. A practitioner would usually want to explore the extent and dynamics of the abusive situation and possibly choose other immediate intervention objectives, such as to have the husband arrested, to refer the husband to a treatment group for batterers, to return the woman to her home, to arrange alcohol or substance abuse treatment for the husband, to arrange marital therapy, to refer the woman to a hospital for a physical examination, and so on. A shelter placement may be the appropriate target in this case, but how could a practitioner possibly determine that without considerable understanding of the nature, dynamics, and history of this woman's experiences in her marital relationship (see Mills 1998, 1999)?

Individual client characteristics must also be considered when selecting targets. For example, the practitioner studying guidelines for the target "finding shelter" would expect different instructions if the client were a homeless man, a thirteen-year-old female runaway, an elderly person with Alzheimer's, or a battered woman. Similarly, under "finding employment," the practice guidelines might need to address different approaches for different clients, such as a high school dropout, a single mother, an AIDS survivor, a alcoholic former executive, or a parolee.

As the early casework scholars demonstrated, a careful investigation of the problem does not require formal classification, but it seems unlikely that practitioners can select targets without first identifying and understanding the nature of the problem(s) and its meaning in the life of the client. Even when no attempt will be made to reverse causal mechanisms, problem formulation is the unavoidable first step in treatment. Undoubtedly, there are clinical examples where appropriate targets might be chosen without problem analysis, but these would be rare instances.

SELECTING INTERVENTIONS WITHOUT IDENTIFYING PROBLEMS

Similarly, it is difficult to find examples of effective intervention without problem identification. We thought we had found one in a recent front-page article in *The Washington Post* (Brown 1999), MEDICINE'S GROWTH CURVE: HEALTHY PATIENTS. The opening sentence is, "What's most noticeable about the steady stream of patients into Charles F. Hoesch's medical office in the Baltimore suburb of Perry Hall is how few are actually sick" (A1). The article describes the rise of the use of medical intervention, particularly pharmacology, in the absence of physical disorders.[4] It features a patient, a sixty-year-old engineer, whose slight elevation in blood pressure would never have provoked notice a generation ago. The patient didn't have heart disease, kidney problems, or any of the other complications of high blood pressure. Nevertheless, he was under medical care. The article's main theme was that the definition of disease has so expanded that it includes people who may be "at risk," which means just about everybody at any age, even if they are currently "healthy." One can be "at risk" even when the probability of getting the disease is slight; the vast majority of those so designated will never get the disease, with or without treatment.

Although this may appear to be a case of treatment without an identified disease, the situation is murky because it resembles many familiar public health efforts to inoculate healthy populations against specific contagious diseases (polio, influenza, etc.) Thus, even treating healthy people rests on some assumptions about preventing possible future diseases.

Social welfare examples of intervention without identified problems are hard to come by, although some social policies and social programs at times appear to be stimulated by vague concerns, promoting nonspecific interventions in order to achieve nonspecific goals. (We would be tempted to use community mental health as an example here.) We could consider these as interventions intended to achieve some general salutary social effects, disconnected from any careful problem analysis. In popular culture, some New Age, feel-good-about-yourself efforts, for example, eating right, exercising, cultivating friends, and pursuing dreams, could be viewed as interventions in the absence of specific identified problems to promote healthy development and raise the general quality of life. We can imagine Quality of Life

---

[4]In our culture's medicalization mania, even the healthy are now sick and treatable, and since drug therapy is the frequent treatment, it is a bonanza for pharmaceutical companies.

Clinics in which clients are helped to develop general life skills and advised about how to make their lives better and more fulfilling, without any attempt to diagnose or identify a particular problem.

Although it may be possible to contrive interventions without problems, for the most part, social work intervention assumes that a client has a problematic condition that is to be altered, prevented, or ameliorated. How that problematic or undesirable state is understood by the practitioner is usually relevant to the selection of the intervention and its objectives.

TARGET-BASED CLASSIFICATION

Let's now consider the scholar's task of organizing knowledge for use by practitioners. Currently, social work has some knowledge organized by general intervention technique (e.g., skills training, cognitive-behavioral therapy, casework), some organized by problem (e.g., child abuse, substance abuse, schizophrenia), and not much at all organized by intervention target. Even if there were a valid etiologically based diagnostic system, we could still argue that organizing intervention knowledge by targets would be more useful and more appropriate for practice guidelines, since in social work, distant causes of human anguish are not readily manipulatable by practitioners who are compelled, nonetheless, to offer assistance. For example, the inability to walk can be caused by birth defects, traumatic injuries, or disease, but a social worker can often ignore the etiology and focus instead on acquiring a wheelchair, transportation services, design modifications in the home and office, and so forth. Having guidelines organized under the goal of "increasing physical mobility" could be an advance for the profession in terms of this problem, which currently has wavering schema for intervention knowledge spread across dimensions of problem, setting, field of practice, and a variety of personal variables, such as gender, ethnicity, and age. Organizing intervention knowledge around targets might be an advance that is useful in practice. What are the requirements and potential difficulties for such a taxonomy?

The proposal for a target-based taxonomy is an argument for making intervention objectives the primary dimension for organizing knowledge and practice guidelines. The practitioner should first decide what is to be accomplished (select intervention targets) and then seek knowledge about which type of intervention is most likely to achieve that objective. But, as we have suggested, selecting a target is insufficient; targets and interventions almost always presuppose some prior identification and understanding of the client's problem.

The practitioner also has to consider client characteristics—particularly age and gender—in choosing an intervention strategy (Videka-Sherman 2000). For example, intervention options in cases of domestic violence would be different depending on whether the abused person was a child or a grandparent; job training and placement may present different challenges if the client is eighteen or forty-four years old. Similarly, emotionally troubled teenagers may be vulnerable to different risks depending on whether they are male or female. What this suggests is a knowledge taxonomy that has practice guidelines nested in a multidimensional matrix framed by client problems, intervention targets, and client characteristics. The menu of recommended interventions or guidelines would be listed in each cell.

Currently, as in the American Psychiatric Association's guidelines, knowledge is organized first by disorder (e.g., schizophrenia, cocaine-related disorders) and second within each disorder by intervention technique (e.g., pharmacological treatments, psychosocial treatments), with little said about targets. So Proctor and Rosen's suggestion to use targets as a major dimension for guidelines is radically different, although both schemes require problem classification of some kind. With a problem-by-target-by-client characteristics matrix (as in table 3.2), a practitioner would be directed to the same cluster of guidelines whether he or she began by selecting a target or a client problem.

Challenges in Organizing Knowledge by Treatment Objective

The challenge of an appropriate taxonomy does not end with the labeling of target categories. A classification system organizing intervention

TABLE 3.2 Practice Guidelines by Problem, Target, and Client Demographics

PROBLEMS	TARGETS/OBJECTIVE		
	Improve Parenting Skills Client Demo A-Z	*Safe Environment* Client Demo A-Z	*Treat Injuries* Client Demo A-Z
A Child Sexual Abuse	Guidelines A1	Guidelines A2	Guidelines A3
B Child Malnutrition	Guidelines B1	Guidelines B2	Guidelines B3
C Child Neglect	Guidelines C1	Guidelines C2	Guidelines C3

knowledge by targets must meet the same requirements as any other system. The target categories need to be meaningful (valid) and the system must be used consistently (reliably). Since we are addressing only the general idea for such an organization rather than a specific proposal, we can only speculate about what challenges might be encountered.

VALIDITY

In terms of validity, there will undoubtedly be much work to do in defining the meanings of target terms. Let's examine the earlier example of finding shelter for an abused spouse. How did the practitioner define the intervention objective? Does "finding shelter" mean separation from the abuser, placement in a safe environment, having food and a place to sleep, acquiring transitional housing, or getting help in terminating an intimate relationship? With each definition, a slightly different body of knowledge would be accessed and a different set of guidelines might be found. For example, if the goal were to end a relationship, the guidelines might be anchored in knowledge of separation and divorce. On the other hand, if the goal were transitional housing, the worker would look at some index term for temporary housing to find guidelines about best placement practices. Organizing knowledge regarding placements of this type would require considerable research to clarify how that target is different from other out-of-home transitional placements for adults, including adult foster care, board and care homes for those with serious mental illness or Alzheimer's, nursing homes, halfway houses for parolees, residential treatment facilities for substance abusers, etc. Some of these placements would undoubtedly share practice guidelines, but determining which ones are similar in consequential ways would take a lot of research effort.[5] The definitions of all the common intervention targets would have to be scrutinized in this way and the overlaps and conceptual connections and distinctions among them hammered out. This difficult, labor-intensive scholarship will occupy a platoon of practice researchers.

PRACTICAL USE AND RELIABILITY

The developers of *DSM* and PIE know first hand the enormous challenge in getting committees of experts to agree on the meanings of cate-

---

[5] Similarly, client characteristics such as gender, age, and ethnicity may interact with these best practices and require further refinement (see Videka-Sherman 2000).

gories and to use them consistently. Even after the experts achieve agreement on the definition of targets, there is always the question of whether practitioners facing the same client or picking the same general target will use the matrix and the guidelines reliably. It is one thing to develop a tidy classification system on paper and quite another to have actual practitioners use it consistently in the way it was intended. This problem has spawned a small field of clinical study about the differences between "efficacy" in a controlled research setting and "effectiveness" in the real world, where clinical discretion, self-selection, organizational pressure, bias, and misuse occur (Seligman 1995). The reliable use of guidelines, whether organized by problems or targets, will be an enormous challenge.

OTHER ISSUES

There are other potential problems familiar in social work, such as the multiproblem client or in this case, the multitarget client. The abused spouse may need more than a safe environment for a few days. She may need permanent alternative housing, a job, child care, and skills to manage the abusive partner, and each target directs the practitioner to a different practice guideline. Will there be a hierarchy of targets or will they be treated equally? In addition, there is the common mistake of trying to fashion a solution from a list of targets without taking the broader context of the client's life into account. This is a fundamental flaw in *DSM*'s descriptive classification, where the symptoms of mental disorder could be indicators of adaptive functioning rather than of pathology if the social context were permitted inside the diagnostic tent (Kirk et al. 1999). Goals and interventions must be sensitive to the contours of the client's life and history and can fall short if the social worker has insufficient understanding of what the client needs. These are not new problems created by the proposal for target-based intervention; they are nettlesome issues as old as social work.

Finally, target-based classification cannot supplant problem classification in one specific arena: the search for explanations of the causes of problems. (We are not assuming that this is what target-based classifications propose.) Assessment of the value of any classification system has to be in terms of its purposes. In the search for etiology, whether of AIDS, child maltreatment, Alzheimer's, or violent behavior, there are enormous scientific efficiencies to be gained by systematically defining the problematic condition. Doing so permits replication by the same researcher and by others at different sites, so that findings are comparable and accumulative regarding correlates and possible contributing causes. This search may not lead immediately to effective interventions, but it will lead to the accumulation of basic knowledge. Moreover, without an established method of identifying problem conditions, it would be difficult, if not impossible, to replicate intervention research or to know the limits of a treatment's generalizability. In the search for knowledge through research, problem classification is indispensable. Organizing knowledge for intervention is different from organizing it for understanding. We need not disparage or dismiss one purpose in order to pursue the other.

Conclusion

Social workers have long been concerned about both the iatrogenic effects of labeling—the stigma that can be associated with being categorized as poor and unworthy, delinquent, or insane—and the obstacles that labels can create for rehabilitation. Yet categorization is a fundamental part of our cognitive ability to organize our experiences and perceptions and to study our world. The suggestion that we organize practice knowledge and guidelines in terms of target-based outcomes is more than a reaction to negative labeling. It is a reaction to the professional penchant for peering into a client's biographical past for tentative, and at times dubious, explanations for personal problems rather than preparing a plan for effective intervention. Although early casework theorists emphasized diagnosis as a basis for determining action, there was always the danger that the search for the "true" diagnosis would become an obsessive end in itself. Thus, the proposal to shift the organizing rubric for intervention knowledge from client problems to targets is intriguing. It has the advantage of being action-oriented, tied directly to the practitioner's immediate decisions, and prescriptive of how the current empirical literature might be reorganized.

This shift, however, will not obviate the need for problem analysis. Practitioners will always have to attend to a client's discomfort, dig into its origins, map its contours, and tentatively formulate some understanding of its nature *before* selecting a target of intervention. And researchers will still have to struggle with the subtle details of problem categorization in the search for contributing causes and common dynamics that may eventually lead to effective treatment or prevention. But practitioners and researchers do not necessarily need the same level of problem diagnosis or classification. Their work serves different purposes, and to assume that one formal classification system will be good for both, as the promoters of *DSM* did, may lead to a confused system that serves neither well.

For practitioners, it is perfectly sensible to formulate a tentative understanding of the client's problem before deciding what they want to achieve and how to achieve it. Nevertheless, while problem assessment is primary, it does not require a formal classification system sprinkled with awkward category labels and constantly changing criteria and encrusted with administrative code numbers. Merely to distinguish, in the words of Helen Perlman, among problems "of wanting a child and of wanting to get rid of a child, of needing money and of wasting money, of not wanting to live and of not wanting to die" may be enough. Any reorganization of social work knowledge should acknowledge our need to understand clients' problems as we develop plans to help them.

CHAPTER FOUR

The Scientific Model in Practice: The 1960s and Beyond

A s we discussed in previous chapters, the scientific method has been viewed as a model for the assessment and treatment of individual cases since the earliest days of the profession. As it evolved, the model called for practice to be scientific in the sense of being a rational, systematic, problem-solving activity. This description is of course not unique to science and indeed can be used as a template for practice in many fields not usually thought of as scientific, such as law and journalism. Moreover, social work has tended to be much less systematic and much more subjectively determined than the model would suggest. Nevertheless, the notion that practice could be guided by the scientific method provided a beacon for some practitioners and set the stage for further developments.

The New Look in Scientific Practice

The transformation of this earlier paradigm into a much more elaborate model of scientific practice began in the early 1960s and was led by research-oriented academics who had entered newly developed doctoral programs in schools of social work in the 1950s and early 1960s. As noted, this was a period of rapid development of such programs and the beginning of a major shift in academic leadership from master's- to doctorate-holding faculty.

Many of these faculty had been trained in psychodynamic approaches, which, at the time, dominated direct social work practice. While they may have questioned the weak scientific bases of these approaches, they had little choice but to accept them. Their doctoral training did not provide alternatives but did stimulate additional skepticism. Moreover, it reinforced their research-mindedness, exposed them to developments in the social sciences, and equipped them with tools to study practice. In short, the new leaders' training had prepared them to search for new practice alternatives. Interest in other modes of practice was heightened by the apparent ineffectiveness of psychoanalytically oriented casework (reviewed in chapter 2).

An important nexus of this development was the recently established doctoral program at the Columbia University School of Social Work. Among the students in this program in the late 1950s and early 1960s were a number of future academics who played key or supporting roles in the development of a scientific approach to practice, including Scott Briar, Irwin Epstein, David Fanshel, Trudy Festinger, Harvey Gochros, Henry Miller, Edward Mullen, (Ms.) Ben Orcutt, William Reid, Arthur Schwartz, Richard Stuart, Francis Turner, and Tony Tripodi. A research-oriented faculty stimulated students to think in an empirical and critical way about social work practice. A strong emphasis on social science and research methodology was provided by such faculty members as James Bieri, Robert Bush, Richard Cloward, Alfred Kahn, Lloyd Ohlin, and Herman Stein. Lucille Austin and Florence Hollis brought research perspectives to their teaching of the prevailing psychodynamic approach to practice.

The inadequacy of existing casework methods and the need for a "new wave" that would have a stronger base in research was a recurring topic of discussion. Yet neither faculty nor student research posed any direct challenges to the psychodynamic model or developed alternatives to it. The program prepared students for a different approach to intervention but fell short of helping them develop it.

An alternative that would fit well with the new practice empiricism was, however, on the horizon. In the mid-1960s, faculty and doctoral students at the University of Michigan School of Social Work, under the leadership of Edwin Thomas, started experimenting with the new behavioral methods that had begun to emerge a decade earlier in clinical psychology and psychiatry. In a series of dramatic and controversial presentations at the 1967 Annual Program Meeting of the Council of Social Work Education, this group unveiled their "sociobehavioral" approach to social work (Thomas 1967b).

This and other behavioral approaches carried the notion of framing practice within a scientific model to a far more advanced level. In its most elaborate form, intervention was to be conducted within the context of a controlled research design—the single case or (as it was later called) the single-system design (SSD). The client's problems were to be defined in specific, observable terms usually expressed as the occurrence of some behavioral difficulty. Data on the frequency and severity of the problem over time were to be gathered to provide a baseline prior to the beginning of intervention. Data collection was to proceed in as rigorous a manner as possible, using such measurement techniques as direct observation or standardized instruments. The purpose of the baseline was to determine if predicted changes occurred after intervention was begun. Moreover, the intervention would be manipulated to rule out other factors that might be contributing to client change. For example, it could be started, stopped, and started again to see if having intervention "on" or "off" made a difference in problem occurrence (the "withdrawal/reversal design"). Or clients could be held in preintervention baseline conditions for different lengths of time to see if changes in the problem occurred when intervention began (the "across-clients multiple baseline design"). Measurement of the occurrence of the problem continued during the intervention phase and was graphed to show patterns of change coinciding with the presence or absence of intervention.

This paradigm also encompassed less elaborate use of scientific procedures. Intervention did not necessarily have to be manipulated. Problem occurrence could be tracked over time and inferences made about the possible role of intervention in producing observed changes (the "AB design"). Also, baseline data could be estimated retrospectively, thus avoiding delays in beginning service. But even with such modifications, the model was clearly a much more advanced and complex scientific approach than the notion of practice as a systematic form of problem solving.

This model carried the idea of scientific practice far beyond the conceptions of Mary Richmond or Florence Hollis, and indeed beyond the kind of scientific medical practice previously emulated. Research methods now became a very visible and sometimes intrusive part of practice. However, the model's proponents were quick to justify this as serving the interests of both clients and the profession. Being able to demonstrate to clients that certain interventions resulted in certain changes was a powerful means of enabling them to achieve mastery over their difficulties. The research could also provide ways of testing the effectiveness of intervention methods as a means of knowledge building and of maintaining accountability.

Behavioral approaches to social work soon began to spread from Michigan to other schools. Part of this dissemination occurred through doctoral students and faculty trained in the approach establishing instructional or research activities at other schools. For example, Sheldon Rose, a faculty member who had worked with Thomas at the University of Michigan, took a position at the University of Wisconsin School of Social Work, where he set up a research program in behavioral group work. Doctoral students trained at Michigan in the early years of its behavioral program and the schools of social work/social welfare they became affiliated with included William Butterfield (Washington University); Eileen Gambrill (University of Wisconsin; University of California at Berkeley), and Clayton Shorkey (University of Texas at Austin).

The Columbia students referred to above were also playing important roles in this process. Richard Stuart became one of the leaders of the behavioral movement at Michigan. Impressed with the work of Stuart and others there, Arthur Schwartz set up a behavioral program at the School of Social Service Administration at the University of Chicago. He recruited Elsie Pinkston, a clinical psychologist who had received her training in behavior modification at the University of Kansas, a leading center of this form of practice. Pinkston in turn expanded the Chicago program considerably and in the process influenced the work of William Reid and of his colleague, Laura Epstein, who were developing the empirically oriented task-centered practice model (Reid and Epstein 1972).

Two other Columbia graduates, Scott Briar and Henry Miller, took positions at the University of California at Berkeley. Both had learned something about behavioral methods during their doctoral studies at Columbia and continued to study them. They collaborated on a book that stressed a more empirical orientation to social work practice in general and greater use of behavioral methods in particular (Briar and Miller 1971). Briar subsequently became dean of the School of Social Work at the University of Washington. Under his leadership, faculty with behavioral orientations, including Rona Levy, Steven Schinke, and Cheryl Richey, were recruited and an educational program stressing the integration of practice and research was established. Briar saw the program as a means of training "clinical scientists," practitioners who could use SSD methods to "participate in the discovery, testing, and reporting of more effective ways of helping clients" (1979:132).

However, this new model of scientific practice remained limited to a small number of academic settings and social agencies. It was given a cool reception generally by the social work practice community. A principal reason was that the model was tied to a behavioral approach. Behavior modification was viewed by most practitioners as too circumscribed to serve as a general vehicle for practice. It was interpreted as emphasizing mere "symptom removal" rather than addressing underlying problems. There were concerns about the paradigm being "mechanistic" and "manipulative," making excessive use of external rewards, and downplaying the practitioner-client relationship. Moreover, there was little enthusiasm for using the complex and intrusive methods of the SSD as a part of practice.

Although the proponents of the new model were attracted to behavior modification, they argued that the methods of assessment, case monitoring, and outcome evaluation used in single-system studies of behavioral approaches could be applied to any form of direct social work practice. As Briar commented, "Unfortunately, there has emerged a belief among some social workers that the single-subject methodologies are suitable only to behavioral approaches to intervention. This belief, which has impeded more extensive use of these methodologies, is without foundation" (1979:139). In fact, examples had begun to appear of application of SSD methods to different forms of nonbehavioral practice, including communication (Nelsen 1978) and psychodynamic (Broxmeyer 1978) approaches.

The notion of the "clinical scientist" method (Briar 1979), or what we shall call "scientifically based practice" (SP), incorporated three main ideas: the application of scientific perspectives and methods in practice, including case monitoring and evaluation through single-system designs (SSDs); the development of new knowledge by scientist-practitioners using SSDs and other research designs; the application, to the extent possible, of interventions whose efficacy has been demonstrated through research. In this chapter we focus on the first two components, which are further elaborated on in the two chapters following. The third component will be taken up in chapter 7.

AN EDUCATIONAL STRATEGY

Although scientifically based practice (SP) was meant to be used by social work practitioners, wholesale adoption did not seem in the offing. Even if it could apply to all forms of practice, the time and effort required, the intrusiveness of SSD methods, and practitioners' lack of familiarity with them were all discouraging factors. The most likely means of dissemination seemed to be social work educational programs, which by now were coming under the control of academics sympathetic to a research-oriented approach.

Beginning in the early 1970s, content on SP, emphasizing single-system designs, was introduced into many research and a few practice courses. Efforts were made to teach SP from both perspectives, through team teaching, integrated research and practice courses, and other devices. As part of their coursework, students were usually required to apply SP methods to one of their own cases in their field placements or employing agencies (Briar 1979; Siegel 1983).

From the faculty's perspective, the courses showed promise of breaking through the legendary antipathy of social work students toward research. Through the single-system design, research could be connected directly to what students were most interested in—practice methods and their own cases—and would have some relevance to their main goal of acquiring practice knowledge and skills. Although some students chafed at what they saw as the intrusiveness of the single-system design in their work with clients, most viewed the courses as, at the very least, a distinct improvement over traditional research teaching. The integrated courses, in particular, had positive effects on the students' attitude toward research, learning research methods, and perceived readiness for practice (Siegel 1983).

As this educational trend gained momentum, its adherents, under Briar's leadership, convinced the Board of the Council of Social Work Education to require social work education programs, both graduate and undergraduate, to prepare students to evaluate their own practice. The Council's action (in 1984) provided further stimulus to the movement. By the late 1980s, a comprehensive survey revealed that a third of the graduate schools of social work emphasized single-system designs and practice self-evaluation in their research; additional schools had substantial content of this kind in their curricula (Fraser, Lewis, and Norman 1991). Integrated formats for teaching practice and research were reported by almost 40 percent of the schools.

How successful has this educational strategy been in disseminating SP? Since the early 1980s, a number of studies have attempted to provide some answers (Gingerich 1984; Richey, Blythe, and Berlin 1987; Kirk and Penka 1992; Marino, Green, and Young 1998; Millstein, Regan, and Reinherz 1990; Mullen and Bacon 2000; Penka and Kirk 1991). (It is worth noting that the data for all but the two most recently published of these studies were collected during the 1980s. Since then interest in the topic has apparently diminished.) Samples of graduates of particular programs or of practitioners in particular agencies (or in general) have been asked to respond to questionnaires about the extent to which they have used SP. The studies have reported substantial use of such components as specifying target problems and goals, describing goals in measurable terms, and monitoring client change over time. Only small minorities of subjects have reported much use of more time-consuming or intrusive operations, such as standardized instruments or graphs to measure change, although the more recent studies show a higher use of standardized instruments (chapter 9). Also, only small minorities of subjects—about 10 percent in several studies—reported conducting single-system studies relating to their practice. Lack of time and agency support as well as interference with practice have been cited as major reasons for not using research methods in their cases. Tolson (1990) and Kagle (1982) have noted as obstacles the turbulence of many settings and the brief, crisis nature of much social work practice.

As has been suggested by Briar (1990) and others, these results can be interpreted as evidence that the educational programs in SP are having some impact. A number of SP components are being widely used, and at least some practitioners are involved in research. Further, the amount of exposure to SP content in graduate school appears to be correlated with the extent to which they use these components (Kirk and Penka 1992). However, these results must be interpreted with some reservations. As Richey, Blythe, and Berlin have pointed out, "semantic differences in descriptions and definitions of component evaluation activities may result in the underreporting and overreporting of such activities" (1987:18). For example, Penka and Kirk (1991) found in their study of NASW members, two thirds of whom had had little or no exposure to single-system designs in their graduate education, that subjects reported using such components as "operationalizing target problems" or "monitoring client change" with more than three quarters of their clients, on average. Many of the practitioners in the studies may be interpreting such components in ways that would be hard to reconcile with conceptions of SP. For example, "monitoring client change" could be understood as simply asking clients about their progress in the course of a clinical interview. Some of the practitioners may be using the kind of scientific orientation advocated by Richmond and Hollis, not unique to SP. To what extent are their responses influenced by a wish to appear "scientific," especially on a questionnaire sent to them by researchers?

In a sense, the subjects may be hitting the semantic ball back into the empiricists' court, forcing them to come up with more discriminating measures of SP components. There is also need for more discriminating studies of practice that might examine actual samples to give more satisfactory answers to some of the questions raised. Finally, studies show little use of more time-consuming but potentially useful tools, such as standardized instruments, despite the availability of comprehensive packages developed by academic researchers (see, for example, Hudson 1982; Fischer and Corcoran 1994).

AGENCY PROJECTS

Another means of disseminating SP approaches has been projects in which training as well as instruments and other materials are provided to practitioners in agencies, usually by academic researchers (see, for example, Mutschler 1984; Mutschler and Jayaratne 1993, Toseland and Reid 1985; Kazi and Wilson 1996a, 1996b; Kazi, Mantysaari, and Rostilla 1997). The most extensive projects have been carried out by Kazi and his colleagues in the UK and Finland (Kazi, Mantysaari, and Rostilla 1997). These and their predecessors have been successful in getting practitioners to use elements of SSD methodology, such as putting targets of intervention into operational form, collecting baseline (often retrospective) data, and using data for decision-making and evaluation purposes. The more time-consuming and demanding the procedure, the less likely it is to be used; little use of controlled designs is reported.

The extent to which SSDs have been used postproject has varied from little or no use to a high level of continued implementation. A key variable appears to be agency support. For example, in one project (in Finland), practitioners, left to their own devices, had stopped using the designs by the six-months follow-up (Kazi, Mantysaari, and Rostilla 1997), but in another, at the Kirlees Education Social Work Services, single-case evaluation became a standard feature of agency practice (Kazi and Wilson 1996a). Despite their limitations, it appears that initiatives of this kind may help SP become established in supportive agency settings.

The Scientific Practitioner as a Producer of Knowledge

The reviews above have suggested some use of SSDs in agency practice. Even if this involved only a small fraction of practitioners, it could conceivably result in a significant number of studies. However, if a study is thought of as producing knowledge for the profession, then it should be published in some form, typically in a journal.

How much and what kind of published research has been produced by the scientifically based practitioner depends entirely on how that term is defined. The original notion of the clinical scientist developed by Briar was a "direct service practitioner" who "participates in the discovery, testing and reporting of more effective ways of helping clients" and "communicates the results of his or her evaluations of practice to others" (1979:132–133). The SSD was seen as a natural vehicle for the scientist-practitioner's research efforts. If this definition is interpreted, as it generally has been, to mean that the clinical scientist is a full-time, agency-based practitioner, there have been very few engaged in doing and publishing research, whether using single-system or group designs.

What emerged was a different cadre of clinical scientists consisting of academic practitioner-researchers, their doctoral students, and collaborating agency practitioners. Leadership was usually provided by the academic practitioner-researcher, who typically had prior agency experience in direct practice and had completed the doctorate. Many taught practice and continued to do or supervise clinical work. This new breed of faculty member—many of whom were trained at or became affiliated with the schools of social work referred to earlier—challenged the existing model in which "practice faculty" were sharply distinguished from "research faculty."

These practice-research faculty members attracted doctoral students who themselves had practice backgrounds, and also initiated research projects in agencies, sometimes in collaboration with their students. The result was a growing body of intervention research. Some of it made use of SSD methods, following the early notion of the natural fit between that methodology and the work of the scientist-practitioner. The first studies began to appear in the late 1960s (e.g., Stuart 1968; Sundel, Butterfield, and Geis 1969), and by 1980 more than twenty-five by social work researchers had appeared in the literature (Thyer and Thyer 1992). This level of output—on the order of two to four published studies a year—has continued to the present. For example, in his comprehensive meta-analytic review of social work effectiveness studies, Gorey (1996) found nine published SSDs from 1990 to 1994.

These single-system experiments have been completed primarily by academic practitioner-researchers and their students. A small number, for example, Edwin Thomas (University of Michigan), Elsie Pinkston (University of Chicago), and Bruce Thyer (University of Georgia), have accounted for a substantial proportion of the output. It should also be noted that social workers have apparently produced more literature on SSD methodology, issues, and the like than on studies that actually used SSDs, to judge from Thyer and Thyer's (1992) comprehensive bibliography of SSD literature.

The contribution of these single-system studies to the development of social work intervention knowledge is difficult to appraise. They certainly have provided evidence supporting the efficacy of a number of (mostly behavioral) interventions for such problems as phobias, marital discord, child adjustment, and difficulties of the frail elderly. However, most of the studies have been one-shot affairs. There has been little use of replication series (Barlow, Hayes, and Nelson 1984) as a means of establishing generality of results. That many of the studies have been published in nonsocial work journals has also limited their impact on social work practice.

A more influential form of research has consisted of group experimental tests of programs that academic practitioner-researchers have designed and directed. These represented a major departure from the types of experiments dominant in social work research prior to the 1970s (reviewed in chapter 2), and were exemplified by *Girls at Vocational High* (Meyer, Borgatta, and Jones 1965), in which, as we noted, researchers were cast primarily in the role of evaluators with little involvement in the design and operation of the service programs. Indeed, since the early 1970s the majority of service experiments carried out by social workers have followed the practitioner-researcher model (MacDonald, Sheldon, and Gillespie 1992; Reid and Hanrahan 1982; Rubin 1985). Their largely positive outcomes have added considerably to the intervention knowledge bases of social work (chapter 7).

A particular strength of this model is that it enables researchers to design and shape their own interventions together with the means of testing them. Thus, outcome instruments can be better fitted to the expected accomplishments of the program. Moreover, the researchers are perhaps more realistic about what measurable results can be attained from their interventions than were the "program people" who conducted the earlier generation of field experiments. They also favor more structured and more clearly articulated approaches with specific, measurable goals and outcomes, rather than the more diffuse, ambitious interventions tested in earlier experiments. The practitioner-researcher interventions were more likely than earlier, psychoanalytically oriented approaches to follow a scientific model in their implementation (although perhaps less so than interventions carried out as part of an SSD). And these characteristics may have contributed to the positive results they tended to show. A notable limitation, which will be dealt with at greater length in chapter 7, is the "investigator allegiance effect," or the potential for bias that can result from a researcher evaluating an intervention that he or she has had a role in constructing.

In sum, although the agency-based clinical scientist did not materialize, clinical scientists, or practitioner-researchers, did emerge in academic settings. The social agency became a site for their research, with agency staff as collaborators. The practitioner-researchers who used SSDs made full use of a scientific model to implement and test their interventions. Those who opted for group experimental methods still tended to design and test interventions with the kind of structure and systemization characteristic of the scientific method.

ISSUES

The use and advocacy of SP in student and professional work with clients has evoked a good deal of critical reaction. The critics appear to agree on one point: that students and practitioners could better spend their time doing things other than learning about and implementing SP methods. Beyond that, the critics tend to part company, since they represent rather diverse camps. On the one hand are those, such as Heineman (1981, 1994) and Witkin (1991, 1996), who have been critical of the epistemological foundations of mainstream research. For them SP represents an unwarranted extension of a faulty science paradigm into the world of social work practice. On the other hand are those, such as Bronson (1994), Wakefield and Kirk (1996), Rubin and Knox (1996), and Thomas (1978), who seem to have no problems with the epistemological foundations of mainstream research but have objected to SP on a variety of grounds having to do with its application. The issues raised by both camps can be subsumed under two broad questions: Does SP adversely affect services to clients? and Does it add sufficiently to these services, or to their evaluation, to justify the effort put into it?

Potential Adverse Effects. An issue voiced over the years by a number of critics (Bronson 1992; Heineman 1981; Thomas 1978; Wakefield and Kirk 1996) is ethical conflicts in practitioners' use of SSDs as a part of ordinary practice. The battle has been joined chiefly in relation to use of controlled designs. To use such intrusive methods in a practice context may improperly confound research and service objectives. The interests of the client may be subverted to the requirements of the study.

However, this debate has been muted in recent years by the recognition that controlled designs are not often used in ordinary practice. Blythe and Rodgers express this shift: "Over time, we came to realize . . . that many designs, such as withdrawal or multiple baseline designs, rarely were appropriate for practice. . . . [We] began to emphasize simpler designs, such as the AB design" (1993:105). Although even recent texts by SP advocates provide a basis for the continuance of such criticisms—for example, describing controlled designs, along with ways in which they can be used by practitioners, in detail (Bloom, Fischer, and Orme 1999; Blythe, Tripodi, and Briar 1994)—it is clear from the advocates' pronouncements that the use of such designs is a minor and probably expendable ingredient in their current view of SP. Since practitioners are not using the designs anyway, the argument becomes academic, in a quite literal sense.

Perhaps the only remaining aspect of this issue—one that has not been dealt with much in the literature—is the type of baseline used in SSDs. Although advocates of SP allow for reconstructed or retrospective baselines, they favor prospective baselines, in which intervention is delayed until data are collected. In some situations, of course, a prospective baseline is useful for purposes of clinical assessment—for example, it can sometimes help both practitioner and client get a better understanding of the frequency of a problem, as well as immediate antecedents and consequences. Nevertheless, for clients actively seeking help, delaying intervention while baseline data are collected could raise questions about whether service needs are being subordinated to research interests. Given their current tendencies toward retrenchment, most SP advocates may well concede the appropriateness of reconstructed baselines in such cases. However, the issue merits more discussion than it has received.

Theoretical Neutrality. A second issue is the "theoretical neutrality" of SSDs. Advocates take the position that such designs can be used with any kind of practice model. To support their argument, they point to a variety of "nonbehavioral applications," including early examples referred to above, e.g., Nelsen (1978) and Broxmeyer (1978), as well as more recent ones, e.g., psychodynamic practice (Dean and Reinherz 1986) and narrative therapy (Besa 1994). Critics counter that the assumptions or metatheory underlying SP inevitably impart a certain direction to work with clients regardless of the model ostensibly used (Kagle 1992; Wakefield and Kirk 1996; Witkin 1996). For example, problems tend to be seen in terms of specific client behaviors rather than failure of the client's environment or social systems.

The reliance on practitioner-engineered measurement to guide treatment decisions may lessen the client's sense of autonomy; it may empower the practitioner rather than the client. Advocates of SP may be better off abandoning the notion of theoretical neutrality and recognizing that they are operating within a metatheoretical framework, with its own set of assumptions about epistemology and practice. If this framework is made explicit, its rationale can be better articulated.

It may make sense to construe SP as a "perspective" on intervention, comparable to the generalist, feminist, or ecosystems perspectives. Like intervention models, perspectives take theoretical positions and contain practice guidelines. In SP, practitioners are enjoined to collect data on change and to use it for decision making. But unlike models, perspectives are not spelled out in detailed protocols. They are designed to be used in conjunction with more specific intervention approaches, although they may make a difference in how these approaches are used.

Defined as a perspective, with a clear statement of its basic principles and requirements, SP might be used in combination with approaches in which it is not normally employed. For example, at first glance, an SP or empirical perspective may appear incompatible with certain humanistic therapies, since it directs practitioners to help clients define their problems in specific, measurable terms, which might run counter to the manner in which such therapies are often conducted. But there is no inherent reason why specificity, measurement, and other requirements of SP cannot be met without violating the fundamentals of humanistic practice. The application, cited previously, of an SP perspective to a test of narrative therapy (Besa 1994) is a good case in point. Or an SP perspective could be used with participatory research approaches in which practitioners and clients work together on a more egalitarian and collaborative basis than is customary in SP.

This might involve educating clients in empirical methods, as well as being willing to accredit client ideas about measurement, data collection, and the like that might not fit conventional research notions. An SP perspective could be used with advocacy research, even though the practitioner-researcher might need to forgo his or her "neutrality." Thus, in work with a tenants' group, systematic collection of baseline data about housing conditions could serve to demonstrate a need for change. Advocacy interventions designed to improve conditions could be implemented and their effects monitored.

As these examples illustrate, an SP perspective could be combined with a range of approaches that may appear in some ways incompatible with it.

The key to such combinations would be adaptations on both sides. The SP may depart from that used in conventional single-system studies, and its application may differ from the routine. But the result may be a stronger and better-evaluated intervention.

The much-needed development of qualitative methods for SP (Reid and Davis 1987) would facilitate its use with other approaches, especially interventions with complex outcomes that may be difficult to quantify, such as those associated with humanistic practice. Care should be taken, however, to develop and use rigorous methods that can be distinguished from the more impressionistic "qualitative" data collecting and evaluation that is part of standard clinical practice.

Cost-Benefit Considerations. SP involves an expenditure of time and effort on the part of both practitioner and client—in operationalizing targets, collecting baseline data and repeated measures of change, constructing and analyzing graphs, and so on. Is it worth the outlay? Advocates argue that the benefits far outweigh the costs. The procedures enable the practitioner to make data-informed decisions about case planning and help clients understand change processes, and serve purposes of accountability to agency and community. Not necessarily so, say the critics, pointing to the lack of studies demonstrating the effectiveness of SP procedures over traditional methods (Wakefield and Kirk 1996). Moreover, they suggest, data collected may be open to question. Standardized instruments, often developed with samples from the white middle class, may give misleading results with poor and minority clients (Wakefield and Kirk 1996). Graphs of the clients' progress frequently yield ambiguous pictures of change (Ruben and Knox 1996).

Lack of evidence that the use of SSDs does enhance outcome is admittedly a limitation of SP. However, two recent experimental studies have addressed this question. In one of the studies (Slonim-Nevo and Anson 1998), 24 Israeli delinquents who received treatment accompanied by SSD methods (for example, targeting specific problems, collecting baseline data, tracking change through scales, self-monitoring, and graphing) were compared with a group (n = 38) treated similarly but without SSD methods. The SSD group showed statistically better outcomes on self-reports of arrests and school or work participation than the controls at a nine- to twelve-month follow-up. However, no differences between the two groups were found on a variety of standardized scales, including measures of self-esteem, anger, and relationships with parents. Although the two groups were comparable on pretest measures, allocations to the experimental or contrast groups were made by the probation officers who treated them, raising the possibility of selection as an explanation of the findings.

The other study (Faul, McMurty, and Hudson 2001), conducted in South Africa, also compared nonequivalent groups but with a much smaller sample (nine in each group). Clients receiving a standard treatment plus SSD methods showed significantly greater reduction in problem severity than clients receiving only the standard treatment, but potential client and practitioner selection biases seriously limited the validity of the findings. The studies in combination have provided at least some support for the possible benefits of SSD methods and no evidence whatsoever that they detract from service effectiveness.

Pending the results of more rigorous experiments, the outcome-enhancing effects of SSDs remains an open question. Even if effects prove to be minor or undetectable, SSDs can be justified on other grounds, such their role in determining accountability and practitioner self-development (both discussed below).

Accountability. According to its critics, SP even fails on its central claim that it provides a powerful means of establishing professional accountability. This is because the methods customarily used cannot isolate the practitioners' interventions as a cause of whatever changes have been measured. Moreover, accountability is not just a question of determining the effectiveness of whatever methods have been tested. It is also a question of ascertaining if a treatment is appropriate and if it is superior to alternatives (Wakefield and Kirk 1996). If one chooses to step outside the SP metatheory, and there is no one happier to do so than Witkin, then accountability is open to a limitless range of interpretations, including challenging existing forms of oppression on behalf of clients (Witkin 1996).

The critics have added some new dimensions to the accountability question. As they suggest, professional accountability is more than demonstrating practice effectiveness (even when that is possible). It involves showing that the practitioner has used the best possible methods for the case at hand in the most suitable way—that he or she has engaged in "appropriate practice" (Ivanoff, Blythe, and Briar 1997). This larger (albeit fuzzier) concept should replace the one-track idea of accountability as measurement of change.

However, monitoring client change and evaluating it on a case-by-case basis are certainly important aspects of accountability, and they happen to be within our grasp through SSDs, however imperfectly we may attain them. If such data can be collected systematically, they should prove far superior to traditional agency practices of presenting "success stories." They may also satisfy demands of funding and managed care organizations using performance-based evaluation systems for accountability purposes. Evidence reviewed earlier has suggested that agency support is needed if SSD methodology is to be systematically implemented. Providing such support may be increasingly to the agency's advantage in competing for funds in a resource-stingy and evidence-demanding environment.

Benbenishty's (1996, 1997) proposal to recast accountability and SP as agency responsibilities merits consideration in this context. Results from single-system evaluations could be fed into a service-oriented, computerized information system that could not only serve some of the agency's accountability needs but also provide useful feedback to managers and practitioners. Developments in computer-based instrument packages should facilitate this effort (Nurius and Hudson 1993). See also chapter 6.

Student and Practitioner Self-Development. One aspect of the use of SSDs has been glossed over in the debates about their efficacy, contribution to accountability, and so on: their role in student and practitioner self-development. Accurate feedback from their own cases can be an important source of learning for both beginning and experienced social workers. It is reasonable to suppose that SSDs can facilitate self-development by providing case data, although evidence needs to be gathered. Studies of how practitioners actually evaluate their own cases, such as Shaw and Shaw's (1997) qualitative investigation of the kinds of evidence social workers use to determine whether their work has "gone well," are especially needed to optimize the fit between the outputs of SSDs and styles of practitioner self-evaluation.

Conclusions

A major change in scientifically based practice began to develop in the 1960s with the adaptation of methods of single-system study to work with individuals, families, and groups. Teaching students how to use research methods in case monitoring and evaluation has become a well-established part of social work education, and there is evidence of applications of these methods in agency practice.

In SP, as in most movements in social work, accomplishments have fallen short of expectations and are sometimes difficult to discern or accredit. SP's hallmark methodology—the single-system design—has really not become well rooted in agency practice, and there is little reason to assume that occasional use has made much of a difference in agency efforts to evaluate practice or establish accountability. However, it may be used more frequently or facilitate related applications as agencies become more concerned about evaluation in a managed care and performance-based funding environment.

Contrary to the vision of early advocates, there are not significant numbers of agency-based practitioner-researchers who have contributed to intervention knowledge through SSDs. On the other hand, academic practitioner-researchers have made contributions using these designs. Further, the single-system methods, with their bonding of research and practice, may have stimulated the new look in group experimental designs that began in the 1970s, in which academic practitioner-researchers develop and test intervention models.

There is by no means consensus that post-1960s scientifically based practice represents an advance. From its beginnings the movement has been criticized by those who reject the tenets of mainstream research on epistemological grounds. More recently, and perhaps more seriously, some mainstream researchers have called into question its central component—the use of single-system methodology in practice—which they have faulted as a dubious application of unproved benefits and possibly a cause of adverse effects. Although their criticisms may be directed at positions that advocates of SP no longer hold, they nevertheless raise some valid issues concerning evidence for the effectiveness of empirical practice, its presumed theoretical neutrality, and its conception of accountability.

In light of these criticisms and with consideration of other needs of the empirical practice movement, we would recommend that: 1) SP be seen as a perspective, like the generalist or feminist perspectives, that will make a difference on how intervention is carried out, rather than as a theoretically "neutral" methodology; 2) in appraising the benefits of the use of research methods in practice, more stress be placed on their potential value in student and practitioner self-development; 3) conceptions of accountability be broadened to reflect the idea of appropriate practice (Ivanoff, Blythe, and Briar 1997) and not simply demonstrate client change or the effectiveness of whatever interventions happen to be used; 4) there be greater emphasis on connecting single-system methodology to agency concerns with program evaluation and accountability, with aggregation of results of case evaluations into computerized agency information systems (Benbenishty 1996,

1997); 5) efforts be made to develop and apply rigorous qualitative methods within an empirical practice framework; 6) research be conducted to obtain a clearer and updated picture of practitioners' use of single-system methodology in practice and research-based interventions, as well as to ascertain the effects of the use of empirical practice components.

CHAPTER FIVE

Engineering Social Work Intervention

S cience and scientific methods have long played critical roles in the development of the physical technologies we use—airplanes, computers, TVs, pharmaceuticals, and so on. Scientific knowledge gives rise to new technologies and informs their design. The products and devices developed are systematically tested and modified until they meet specified performance standards. In industry, these processes are known as research and development. Converting scientific findings to technology requires the knowledge and skill that is the province of engineers.

If social work can be seen as a social technology, then it may be said that the profession has always sought to make science a guide for development. The scientific philanthropy movement tried to use science to shape the new technology of casework with the poor. The concept of "social engineering," which emerged in the late nineteenth century, has been applied to this and other efforts to apply scientific knowledge to create the means to resolve social problems (Graebner 1987). However, throughout most of its history, social work research and development has lacked specific methodologies or personnel. In short, it has had nothing comparable to the engineering disciplines or to their engineers.

This lack became clear in the failed experiments reviewed in chapter 2. Program design was informed by the practice wisdom of the "program people" rather than by scientific knowledge. Little attention was given to fit between program and research requirements. There was no systematic pilot testing and resultant modification of initial program designs.

Social Workers as Engineers: Early Efforts

Social work engineers of a sort began to appear in the 1960s as the practitioner-researchers graduated from recently established doctoral programs. Combining practice experience with research skills, they sought to invent and try out new service models. They were perhaps able to create better fits between intervention and research components than loose teams of agency social workers and academic researchers. But they still lacked a systematic approach to intervention design, modification, and testing.

Social work and other human service professions began to move toward the development of engineering models. In the climate of the Great Society, social scientists of all types felt pressure to become more involved in applying scientific knowledge to social problems. Accepting the legitimacy of these pressures, Fairweather introduced a model of social innovation in which different interventions for a problem could be compared through a series of experiments (1967:213). As research relevant to the human services began to accumulate, there was growing interest in its utilization by practitioners, including efforts to synthesize findings from the social sciences that might be relevant to practice (see chapter 8; Stein and Cloward 1958; Thomas 1967a). Social R&D, an adaption of industrial R&D, emerged (Guba 1968; Havelock 1968; Rothman 1974). The behavioral model of practice, with its explicit interventions and scientific methodology, lent itself particularly well to an engineering approach (Thomas 1967a).

Engineering models specifically for social work began to be developed in the 1970s. The most substantial work was done by Jack Rothman and Edwin Thomas, colleagues at the University of Michigan School of Social Work. Rothman's social R&D approach adapted the procedures and logic of industrial R&D to social work intervention (1974, 1980). The first step was to identify a practical problem or goal of concern to practitioners. Relevant empirical knowledge was extracted from the literature and synthesized, then used as a base for designing an innovation, usually a type of intervention. The innovation was then operationalized, given guidelines and procedures to be used in carrying it out. Pilot testing was conducted to determine how the innovation worked under actual conditions and to provide feedback that could be used to make further modifications. Main field testing was then carried out in order to ascertain the effectiveness of the innovation and learn more about its performance. The results might lead to further refinements. The final stages of the model consisted of systematic efforts at diffusion, including preparation of detailed guidelines for carrying out the innovation, identifying and recruiting potential users, and assisting them in implementation. Diffusion efforts might lead to further field testing and modifications. Rothman used his R&D model to design, develop, and test a practice model for community intervention (1980) and subsequently approaches to helping homeless and runaway youth (1989) and case management (1992).

Thomas (1978a, 1978b, 1984) also took an engineering approach to the design and production of social technology: select a problem, gather information, design an intervention; develop it through empirical testing and evaluation until it produces the intended results; then take steps to ensure its dissemination. Whereas Rothman focused on adapting industrial R&D to social work purposes, Thomas attempted to create a more general model, which he referred to as "Developmental Research and Utilization (DR&U)" (1978a). Rothman's social R&D called for a review of scientific findings as a basis for the creation of an innovation; Thomas, however, introduced a variety of additional sources, such as technological developments in related fields and new procedures created in practice. He embedded DR&U within a broad framework for developing interventions in social work. For example, he constructed typologies of assessment and intervention methods and devised criteria for determining what is a "good" intervention. He specified a range of social technologies, in addition to intervention methods, for which developmental research might be appropriate, including assessment methods, information systems, organizational structures, and social policies. (However, like Thomas, we will focus on the role of developmental research in the creation of intervention methods.) In general, Thomas (1978a) saw developmental research as perhaps "the single most appropriate model for research in social work because it involves methods directed explicitly toward the analysis, development, and evaluation of the very technical means by which social work objectives are achieved—namely, its social technology" (114). Thomas used elements of DR&U in direct practice projects in the late 1960s and early 1970s. He then applied it more fully in the long-term development of a model for "unilateral family therapy" for wives of alcoholics (Thomas, Santa, Bronson, and Oyserman 1987; Thomas et al. 1990).

Tony Tripodi, also at the University of Michigan School of Social Work, adapted elements of R&D in his model for research, development, training, and evaluation (RDTE) (1974). This approach begins with a series of controlled studies of social work practice. Intervention components with empirical support are identified and built into a training program for practitioners, which is then evaluated. The program is comparable to the kind of intervention technology that would be constructed in the Rothman and Thomas approaches; its training component is analogous to their implementation phases. However, in RDTE the effectiveness of the constructed intervention technology is not tested (presumably having been established in the prior research), but the efficacy of the training itself is evaluated.

Also inspired by industrial R&D, William Reid and Laura Epstein, at the School of Social Service Administration, University of Chicago, embarked on a series of studies aimed at the development of their practice approach, task-centered casework (Reid and Epstein 1972, 1977; Reid 1975, 1978b). They viewed "model-building research" as a tool for the development of this approach rather than as a system for constructing interventions generally. They had already created an intervention based on prior research (Reid and Shyne 1969) and used the findings of their studies as feedback for model development. In other words, they entered the R&D stepwise framework after the initial design was in place but made use of the later phases. Drawing on Thomas's and Rothman's work as it became available, Reid (1985a, 1987) used design and development (D&D) to create methods for task-centered treatment of couples and families. Full-scale application of the D&D paradigm to construct and test a task-centered case management model for children at risk of school failure was carried out with Cynthia Bailey-Dempsey (Reid and Bailey-Dempsey 1994, 1995; Bailey-Dempsey and Reid 1996). Reid also proposed a "model development dissertation," in which a student would construct a tentative practice model, test it through a small-scale "exploratory experiment," and then apply the results to modify and improve the initial model. An early example of such a dissertation was Rooney (1978). More recent examples may be found in Caspi (1995), Donahue (1996), Naleppa (1995), Raushi (1994), and Kilgore (1995).

The approaches outlined thus far have all been directed at development of intervention models for general use. A somewhat different tack was taken by Edward Mullen (1978, 1983), then at Fordham University School of Social Service. Mullen was interested in helping practitioners use research to create their own "personal practice models," which he defined as "explicit conceptual schemes that express an individual social worker's point of view of practice and give orderly direction to work with specific clients" (Mullen 1983:623). The core of such models consisted of "summary generalizations and practice guidelines," which were to be based, to the extent possible, upon available research. Drawing on Rothman's 1974 social R&D approach,

Mullen developed a stepwise protocol for enabling students and practitioners to construct and evaluate their own models, using research, theory, practice wisdom, and other sources.

Despite their variety in form and purpose, these early efforts shared one central and critical feature: they all sought explicit ways of using research to create or improve models of social work intervention. Their aim, as Thomas (1978b) made clear, was to use research to create technology rather than knowledge. For social work this was a novel point of view. The traditional paradigm said, in effect, "We need to do research in order to generate knowledge that social workers can then utilize to improve their practice." The new paradigm said, "We need to make use of research to build effective models of intervention for practitioners to implement." A new role was created, that of the model developer, a practitioner-researcher who would use scientific knowledge to create and perfect intervention methods. Model developers would replace the uncoordinated, and often conflicting, collections of program people and researchers who typically generated and evaluated new interventions. Research utilization was cast in a new light. Practitioners need not necessarily read studies and apply the results; rather, they could use intervention models generated by scientific findings and processes.

Mullen's approach (1978, 1983) appears at first glance to be an exception, since it does set forth a way for practitioners to construct their own models. However, actual applications of his protocol appear to have been limited to special projects or classes. Although in theory practitioners could use the protocols alone, their complexity and time demands suggested the need for a structured, expertly guided experience.

These engineering approaches, especially those of Rothman and Thomas, provided the new model developers with a systematic and detailed blueprint for the production, testing, and dissemination of social work interventions. The steps and procedures were not in themselves foreign to social work researchers, but they had not been previously ordered, explicated, and justified as part of a coherent approach to building technology.

Finally, these approaches were carried forward in social work without the usual heavy borrowing from other fields. To be sure, progress in related disciplines (cited earlier) may have helped launch these developments, but once under way they appeared to be driven largely by their own research activities. Similar efforts were taking place in psychology. For example, Gottman and Markman's (1978) Program Development Model consisted of a set of steps for creating, evaluating, and improving a treatment program.

Although this model was quite similar to those developed by Thomas and Rothman, there was little evidence of mutual influence.

The D&D Paradigm and Intervention Research

Rothman and Thomas had created different approaches that shared many common features. In their writings they regularly drew on each other's work, but they continued on parallel tracks until the mid-1980s, when they collaborated in setting up a NIMH-funded training program in intervention research at the University of Michigan. Then, in 1989, they organized a National Conference on Intervention Research at UCLA, which brought together a group of model developers in social work and related fields as well as experts in research utilization, information retrieval, and technology diffusion. Selected papers from the conference were later published as a book, *Intervention Research: Design and Development for Human Service* (Rothman and Thomas 1994).

In a key paper at the conference (which became the lead chapter in the book), Thomas and Rothman (1994) set forth the concept of design and development, or D&D to integrate the various social work engineering models—social R&D, developmental research, model development, and so on. D&D in turn was seen as part of a larger whole, intervention research, which in their view also included intervention knowledge development (e.g., conventional social science research that might be used to generate interventions) and knowledge utilization (e.g., converting scientific knowledge into intervention guidelines, which would be a part of D&D). In this formulation, knowledge development combines what we have referred to earlier (chapter 2) as assessment knowledge and intervention knowledge.

However, the main focus of the chapter and of the book was D&D, which the authors divided into six stages, integrating their previous efforts. The descriptions of the stages that follow incorporate elaborations made by various model developers, many of whom contributed chapters to *Intervention Research*.

PROBLEM ANALYSIS AND PROJECT PLANNING

A focal point in the first stage, problem analysis and project planning, is the problems or goals for which the intervention is being developed. Information about them may be obtained from a variety of sources, including key informants such as service providers and potential clients (Fawcett et al. 1994).

Once problems or goals have been clarified—or as part of the clarification process—a state-of-the-art review is conducted to determine what work has been done in relation to them. For example, what relevant interventions have been developed? Does the problem even lend itself to social work intervention? The review should help justify the need for the project and provide some initial direction for its subsequent stages. Then a preliminary plan, which specifies the setting, the practitioners, the approximate number of cases, and the project structure is devised and developmental goals are specified.

INFORMATION GATHERING AND SYNTHESIS

In the information-gathering and synthesis stage, an effort is made to acquire and integrate knowledge useful in understanding the problem and designing the intervention. The state-of-the-art review conducted during the previous stage should have begun this task; however, that review was concerned with determining whether or not effective methods for dealing with the problem were already available. What is needed at the second stage is a much more in-depth examination of sources relating to the problem and potential interventions.

A primary source is the research and practice literature, which may be accessed through computerized abstract services such as SWAB (Social Work Abstracts) and PSYCHINFO (Psychological Abstracts) to identify interventions or program components that might be incorporated into the service design, with priority given to interventions with empirical support.

In addition to the literature (and especially when it is in short supply), another information source is study of "natural examples" (Thomas 1984; Fawcett et al. 1994). The investigator can interview practitioners who have used interventions that might be incorporated into the design, or rely on his or her own prior experimentation with the intervention.

DESIGN

Design in D&D refers essentially to the construction of interventions. Although it makes use of previous work, the design phase is basically a creative undertaking in which the developer gives shape to innovative ideas.

Some design activities may involve extending an existing approach to a new population or problem. Others may consist of assembling and adapting an array of interventions from different approaches. Still others may require construction of new interventions for a problem not previously identified or for a problem that has proved resistant to existing methods. To recall a point made earlier, D&D can also include technology other than intervention methods. Mullen, for example, discusses the use of D&D in the design of a computer program to assess a child's risk of harm from a caretaker (1994).

The culmination of this stage is typically the creation of an intervention protocol, which outlines the kinds of problems to be addressed and the assessment and treatment methods to be used. Initially the outline may be skeletal, but it should become detailed and comprehensive as testing of the intervention proceeds. Ultimately, the protocol might include a step-by-step description of procedures to be employed and indications of how different contingencies are to be handled.

EARLY DEVELOPMENT AND PILOT TESTING

As Thomas has observed, "Development is the process by which an innovation is implemented and used on a trial basis, tested for its adequacy, and refined and redesigned as necessary" (1984:169). Early development consists of preliminary trials of the intervention emerging from the design phase. A variety of strategies may be used, from a single trial to successive trials (with the latter preferred). Tests may involve specific components of an intervention model or the entire model. Single-system or group designs may be used. Whatever the strategy, the major purpose is generally twofold, as the quote from Thomas suggests: to determine if the intervention is on the right track and to provide an empirical basis for improving the intervention. A secondary aim of pilot testing is to try out and refine data collection and measurement procedures to devise research methods that are maximally sensitive to the practice operations and outcomes of interest in the intervention.

Being "on the right track" means that the intervention can be successfully implemented and appears capable of achieving its goals. Pilot tests may expose unanticipated feasibility problems with interventions that may have been carefully designed. For example, it may not be possible to obtain a sufficient number of clients or to retain them in the program. Participating service providers, such as teachers, may not cooperate when they realize how much time is required of them. Moreover, the intervention should show some potential for attaining intended outcomes, through measures of client change and other evaluative data. Even though effectiveness cannot be definitively established in a small-scale pilot test, there should be some empirical basis for concluding that the intervention has enough potency to warrant further development.

Most interventions should be able to pass this test, since they usually use approaches that have worked in other contexts. A greater challenge for pilot testing is to obtain data that can be used to improve the intervention, to fill out and correct the rough map laid down in the preliminary formulation. What specific methods were used? Did some work better than others? Were there omissions or shortfalls in the service design? Are practitioners able to perform the suggested tasks? If not, why not? Other questions concern the range and variation of expected events. What kind of case situations were encountered? What methods, of those suggested, were used most frequently, and what did they look like in actual use? Still other questions are related to possible effects of specific procedures. Is the use of an intervention followed by expected changes in the client's behavior? How do practitioners, clients, and collaterals assess the effectiveness of particular interventions? A number of research methods used to answer such questions are summarized below.

Descriptive Study of Intervention. It is important to determine what interventions practitioners actually used to determine how well they followed the service design. The most thorough and accurate methods are tape recording intervention activities and applying some form of content analysis to classify the types of interventions used. Methods requiring fewer resources (and raising fewer feasibility problems) include review (or content analysis) of practitioner logs and narrative records or recording of activities through checklists or other structured instruments.

Critical Incidents. In D&D contexts, critical incidents are typically specific types of events recorded by practitioners (or others) that bear upon how well an intervention is working (Thomas 1984). For example, in a developmental test of a practice model (Procedure for the Assessment and Modification of Behavior in Open Settings), Thomas et al. (1982) collected situations encountered by practitioners that fell outside the scope of their model. In testing their case management model for problems of school failure, Reid and Bailey-Dempsey (1994) defined a critical incident as any instance of parental involvement that appeared to result from the processes of the case management meeting. The collection of critical incidents can help identify program components that need further work as well as those that appear to be successful.

Informative Events. An informative event is an occurrence or episode in a case that inspires useful new thinking about ways to improve the intervention (Davis and Reid 1988; Reid 1985a). It can be seen as a qualitative form of critical incident. Whereas critical incidents are defined in advance of data collection and their occurrences quantified, informative events of various types are identified and collected in the course of reviewing written or taped case material, with less emphasis on quantification. Informative events generally indicate promising techniques, inadequacies of service design, and instructive successes or failures. For example, in Naleppa and Reid's (1998) pilot test of a case management model for frail elderly in the community, one of the informative events identified the service design's lack of methods of dealing with clients' tendencies to reminisce during interviews. The protocol was revised to provide ways for the client to reminisce while still maintaining focus on the target problem. Identification of even a single occurrence of an event can occasion a revision in the intervention.

Interpersonal Process Recall (IPR). In IPR the practitioner and client together review taped selections of a treatment session (Elliott 1984). Clients may be asked to react to key events that they or the practitioners select. For example, Naleppa and Reid (1998) asked frail elderly clients to comment on interventions used. Was the practitioner being helpful at this point? What could he or she have done to be more helpful?

Task Analysis. This method is an attempt to determine the sequence of steps clients or practitioners can use in working through a particular problem. As Thomas (1984) has suggested, it is useful to distinguish between rational and empirical task analysis. The former is based on prior knowledge and logical considerations. No data are collected. The latter is based on the collection and analysis of data—for example, how practitioners help pregnant teenagers reach a decision about whether to keep or surrender their expected children.

Rational and empirical task analyses can be used together. For example, empirical analysis of how clients actually proceed with an issue can be combined with a rational analysis of possible options to produce an ideal model of how the client can solve the problem. This kind of integrated task analysis was used by Berlin, Mann, and Grossman (1991) to construct a model of how depressed clients could resolve dysfunctional appraisals by significant others. Such models can then be used to design practitioner interventions. (See also Greenberg 1984.)

Analysis of Interventions and Proximal Outcomes. In this procedure, the outcomes of specific interventions are examined immediately or at regular intervals during the course of treatment, using such data as tapes of treatment sessions, practitioner recording, or client responses to questionnaires given at the end of the session (Orlinsky and Howard 1986). The purpose is to determine if particular kinds of interventions are more likely than other kinds to be followed by predicted short-term changes. For example, Reid (1994b) found that family members were more likely to carry out tasks at home if they had discussed and planned the tasks together in a structured problem-solving sequence rather than simply developing them in an unstructured discussion with the practitioner.

Other Feedback from Practitioners and Clients. Feedback from practitioners and clients may be a part of each of the methods described above. Additional information can be obtained from practitioners through conferences or supervisory sessions during the pilot test (Rothman and Tumblin 1994). Postservice questionnaires, interviews, or focus groups can be used with practitioners and clients to acquire retrospective data on intervention processes and outcomes, with emphasis on how well different components worked.

Quantitative and Qualitative Methods. As the examples above suggest, methods used in early development (or later) to improve the intervention are a mixture of quantitative and qualitative techniques. The latter are used particularly to develop themes from data yielded by critical incidents, informative events, interpersonal process recall, and task analysis. Grounded theory methods (Strauss and Corbin 1990) may be useful in identifying patterns and hypotheses. Because quantitative data may be inconclusive with small samples, they may often be best used to supplement the qualitative data. In D&D this mix serves a critical generative function. The developer is not only interested in identifying patterns and forming conclusions about intervention processes but also open to ideas about the intervention that might be trigged by immersion in the data. Thus the data are used to stimulate creativity. For example, in conducting a task analysis, a developer may begin to tease out the steps a practitioner used to achieve a goal with a client. In the process, he or she might see a better way to achieve the goal. This "better way," rather than the practitioner's actual behaviors, might then serve as a basis for improving the model. Similarly, analysis of an informative event—say, a practitioner making effective use of a novel technique—might stimulate ideas about how this technique could be used in other situations (Reid and Davis 1987; Davis and Reid 1988). Its broader use might then be incorporated into the intervention model. Such modifications may not be empirically based in the sense of being grounded in data providing unequivocal support for their efficacy. But they are so in the sense that immersion in the data may be necessary for the developer to conceive of them.

EVALUATION AND ADVANCED DEVELOPMENT

In this stage of the D&D paradigm, the intervention is put through a more rigorous test of effectiveness. Ideally, some form of controlled experiment is used to determine if the intervention has produced expected changes. In addition, developers learn about the relative effectiveness of different components of the intervention, the relative effectiveness of the intervention compared with other interventions, and how efficient it may be (by conducting cost-effectiveness and cost-benefit analyses) (Thomas 1994). Additional pilot testing may be employed to see how the intervention works under "normal operating conditions" (Thomas 1994:194).

The evaluation phase usually produces findings that may lead to further refinements in the intervention. Also, developmental goals can be built into the evaluation design—for example, alternative components of the intervention can be compared. However, advanced development is limited to refinements of the original design; it does not embark on significant new constructions.

DISSEMINATION

The sixth and final stage, dissemination, occurs (ideally) after the intervention has been rigorously tested in the previous stage. As suggested earlier, dissemination in D&D requires model developers' active effort to bring about the actual implementation of their innovations. To this end, they may attempt to "market" the intervention in a manner similar to the marketing of industrial products (Corrigan, MacKain, and Liberman 1994; Fawcett et al. 1994). Marketing may include active "selling" to agency directors and their staffs and developing strategies for overcoming administrative and staff resistance to using it (Corrigan, MacKain, and Liberman 1994).

Moreover, the D&D paradigm spells out procedures and strategies for dissemination. The product needs to be prepared for use. Intervention guidelines, instructions, etc. often take the form of detailed protocols or manuals; model developers try to put them into the hands of potential users and facilitate implementation. The preparation of conference papers and journal articles about the innovation may be secondary efforts at dissemination.

Variations and Elaborations

The paradigm presents a linear model for the development of interventions from scratch with a problem as the starting point. That variations on and elaborations of this paradigm are possible is well illustrated by the long-term D&D effort conducted in relation to the task-centered practice model (Reid and Epstein 1977; Reid 1978b, 1985a). As noted earlier, the developmental process began with an intervention method in hand. As a result, the problem analysis and project planning stage was concerned with applying the model to particular types of problems. Information retrieval and synthesis was directed at learning more about these problems, the kinds of remedial interventions that might be incorporated within a task-centered approach, and alternative forms of intervention. This type of development might be characteristic of any effort to use research to improve an existing mode of practice. The developers might not be interested in perfecting interventions for given problems but rather in determining how a particular intervention might be best modified for a range of problems. Similarly, the invention of almost any hard technology, such as the harnessing of electric current or the discovery of x-rays, is followed by efforts to find new applications. Much of the research on established modes of intervention in the human services, such as applications of behavioral and cognitive treatment for various problems, has taken this route, even though D&D principles may not have been explicitly used.

Moreover, developmental work with the task-centered model has taken place over a thirty-year period and is continuing. This effort has involved much more than the kind of project set forth in the D&D paradigm. There have been a number of controlled evaluations of variations of the model for different populations and problems (see Reid 1997c for a review). Some cases revealed that while the model was effective overall, there was considerable room for improvement with certain types of clients or problems, which led to further development and testing (see Reid et al. 1980 for an example). In one instance (task-centered case management with children at risk of school failure), the controlled evaluation failed to produce significant results and in retrospect appeared to have been premature (Bailey-Dempsey and Reid 1996). Prior pilot testing had been inadequate. The failed experiment was used as a pilot test: process and outcome data were extracted and techniques described previously for use in early development were applied in an effort to identify reasons for the failure and ways of improving the model. The model was then retested in another controlled evaluation with positive results (Reid and Bailey-Dempsey 1995). As these examples suggest, there is need for expansion of the D&D paradigm itself to encompass a broader range of routes and contingencies in developmental processes.

Issues

What have these engineering approaches contributed to the development of social work knowledge? What issues have they raised? In considering such questions, we shall let the D&D paradigm stand for related and predecessor approaches—developmental research, social R&D, etc.

OBSTACLES TO USE OF D&D

D&D has provided a rational framework as well as a systematic set of procedures for creating, testing, modifying, and disseminating social work technology, but its influence is difficult to determine. Explicit uses of D&D (beyond the small number of projects already cited) have been limited, although instances can certainly be found (see, for example, Dore [1999] and Icard, Schilling, and El-Bassel [1995]). However, research efforts in social work have frequently followed the spirit of D&D in one way or another: building innovations from previous work, using pilot studies to inform the development of intervention models that are then tested more rigorously, trying to disseminate programs, and so on. But references to

D&D are the exception rather than the rule, and there has not been much recent work in related fields using D&D-type frameworks.

A reason for this is that D&D components reflect a common-sense approach for practitioner-researchers intent on building and testing intervention models. Once social work practitioner-researchers took on engineering roles, they began to behave like engineers, accepting extensive literature reviews, pilot studies, etc. as logical methods to use in a scientifically oriented, systematic approach. However, these methods have not been used routinely or with the kind of thoroughness called for in the D&D paradigm. Although the methods show the influence of D&D literature, this influence is not made explicit in reports of most recent experiments.

A major obstacle to extensive use of D&D approaches (whether implicit or explicit) is the expense and time required to carry them out. Funding agencies are typically more interested in supporting definitive experimental tests of fully formed interventions than "preliminary" developmental work. In fact, they may expect the latter to be done prior to the submission of a grant proposal. Because funding may be difficult to procure, the initial stages of the D&D process may be given short shrift. Reviews of prior research may be sketchy and pilot testing may be quite limited or omitted altogether. Pressures from university administrators for faculty to engage in externally funded research and time constraints in meeting proposal deadlines are additional obstacles to use of D&D.

THE MISSING LINK IN EARLY DEVELOPMENT

A critical stage in D&D is the pilot testing phase in early development, through which the model developer hopes to gather data that can be used to identify difficulties in the intervention and to form a basis for improving it. In the experience of one of the authors (Reid), early pilot testing is most useful in identifying obvious shortcomings in the intervention and generating ideas about how it might be improved. It is less useful in determining the relative effectiveness of different components that show varying results. To get hard evidence that component A is more effective than component B (as a basis for a decision about whether to keep A and drop B) would necessitate using complex controlled designs, which may be impractical at this stage. Without such designs, early pilot testing involves a considerable degree of guesswork. In industrial research and testing, whatever controls may be needed to determine if component A works better than B are much easier to come by than in tests of social work intervention. Here especially, the analogy between industrial R&D and social work D&D breaks down. It is doubtful that the appropriate research methods can be developed in social work to close the gap. Still, informed guesses based on systematic data may provide a better basis for constructing interventions than even more impressionistic alternatives.

SELLING THE PRODUCT

In D&D a concerted effort is made to promote the implementation of tested intervention models. This is in contrast to the usual practice in conventional intervention research, in which findings are offered to the scientific and practice communities through publications, conference presentations, and the like but without a sustained effort to convince others to adopt the interventions. The rationale in D&D is that effective interventions are more likely to be used if their developers actively promote them. Moreover, the involvement of model developers in facilitating use of their products can help ensure that they are implemented in the manner intended. Finally, there are ample precedents for the active involvement of researchers in "selling" their findings. In the social survey movement discussed in chapter 2, investigators undertook studies with the intent of using findings about social problems to persuade community decision makers to take action to resolve them. In present times, researchers using action research models, among other approaches, employ similar strategies to bring about implementation of their own findings.

Nevertheless, there are risks in practitioner-researchers aggressively selling their own wares. Investigator allegiance is a likely source of bias in any undertaking in which a researcher evaluates his or her own brainchild (see chapter 7). A researcher's considerable investment in an innovation could introduce tendencies, blatant to subtle, to inflate its effectiveness. Such biases are not likely to come to light when the researcher becomes a salesperson. Moreover, the anticipation of marketing a product, with the possibility of gaining fame and fortune as a result, could further accentuate investigator bias.

WHEN IS IT READY TO GO?

The D&D paradigm calls for dissemination following rigorous evaluation that establishes the effectiveness of the intervention. This is an ideal that model developers should strive for. Still, dissemination without such testing can be justified if the intervention shows promise (Curtis 1996) and if there is nothing in place that has been shown to be more effective. Models of intervention that have completed the steps of the paradigm prior to evaluation have been systematically designed, tested in a field trial, and redesigned in light of the test. They have attained a measure of "developmental validity," which refers to "the extent to which interventions have been adequately used on a trial basis and have been tested developmentally" (Thomas 1985:54).

If the model is intended to meet a need in practice that is not currently addressed by interventions with greater developmental validity, then there is justification for its implementation in practice and even its dissemination, in the absence of a controlled experiment. Because D&D is generally not used if tested methods are already available, interventions shaped by the D&D process generally are better tested (i.e., have greater developmental validity) than alternatives. If the new interventions prove promising, they should be tested more rigorously, but even pending such tests, they should take precedence over counterparts that have not gone through a developmental process. Moreover, as Curtis (1996) points out, dissemination of preliminary pilot tests can stimulate others "to try new ideas and to add their insights into the investigation" (118).

Another consideration in the evaluation and dissemination of a D&D product is prior evidence on its effectiveness, perhaps obtained in studies of its use with other populations or problems. Most D&D undertakings build on previous efforts. Some may involve variations of methods found to be effective in earlier studies. For example, an anger management program may have been found to be effective with mildly retarded adolescents between the ages of thirteen and sixteen. A similar program is developed for mildly retarded children ages nine to twelve and tested in a field trial without a control group. The pre-post changes for the younger children resemble those for the adolescents treated and surpass those for the controls used in the evaluation of the adolescent program. Such evidence could be used to argue that the program is probably effective with younger children as well as adolescents and would strengthen the case for dissemination in the absence of programs with better empirical credentials. Similarly, applications of the task-centered model of social work intervention have been disseminated after initial field trials but prior to controlled evaluations (see, for example, Rooney's [1981] model for helping parents regain their children from foster care and Naleppa and Reid's [1998] case management approach to working with frail elderly in the community). Although the effectiveness of these applications cannot be firmly established, they gain credibility from a number of controlled studies that have demonstrated the effectiveness of task-centered practice with a variety of populations and problems (Reid 1997c).

Just as a case can be made for an intervention's dissemination prior to formal evaluation by the developers, a case can also be made for more caution prior to replication by investigators with less investment in the product. In an ideal world, an independent replication could well be considered essential prior to any dissemination, as is presently the case with therapeutic drugs. However, such a position would be excessively limiting in the human services, where replications are rare and the need for interventions with at least some empirical credentials is compelling. Still, we may be wise to be less assertive in "selling" unreplicated interventions.

The direction suggested by Curtis (1996) as well as by Chambless and Hollon (1998) may provide a solution: to develop criteria for efficacy that can be used to rate interventions. Using Curtis's (1996) criteria, we have seen that an intervention that shows positive results in uncontrolled field testing can be considered "promising." If the results hold up in a controlled evaluation, the intervention, according to Curtis, can be considered "probably effective." Several controlled studies with positive results would be needed in order to designate the intervention's effectiveness as "established." Chambless and Hollon (1998), following recommendations by a task force of the American Psychological Association, developed somewhat stricter criteria. Their first rank goes to interventions that are "efficacious and specific"—whose efficacy has been demonstrated in at least two independent replications; also, there must be evidence that they have surpassed either a "placebo" condition that has controlled for nonspecific effects or a rival treatment. Interventions that have been shown to work in two independent replications but have not demonstrated superiority over a placebo condition or a rival intervention are considered simply "efficacious." If an intervention has had only one independent replication or been conducted entirely by one research team, it is "possibly efficacious." By these standards, interventions that have been demonstrated to be effective in the evaluation called for by the D&D paradigm would qualify as either "probably effective" (Curtis [1996]) or "possibly efficacious" (Chambless and Hollon [1998]).

Such standards could be used to indicate the level of confidence to be placed in a disseminated intervention. Although an intervention considered promising or better might be disseminated, its "confidence rating" would provide some guidance about how hard it should be "sold" to potential users and the caution they should take in implementing it.

Conclusion

The D&D paradigm has presented a systematic approach to the creation of empirically tested intervention methods in social work. Although the paradigm has not yet had a pervasive or obvious influence, it has produced some well-developed examples of how it might work and presumably has reinforced trends toward greater coherence in the design and testing of social work interventions.

At a fundamental level, work on D&D has addressed basic issues of how social work can develop better means of creating effective technology. Is it possible to devise methods for engineering the construction of social work intervention that parallel well-established methods of engineering hard technology? When we go beyond the tactics used in designing or testing an intervention in a given experiment to principles of the design and testing of interventions in general, we are in D&D territory even if we do not use its current terminology.

At this level a number of key questions occur. Are there preferred general strategies of searching the literature for available interventions that might be used in building a new model? What kinds of designs and samples make the most sense in pilot tests of an innovation? How can research be best used to improve practice approaches? What kinds of empirical verifications do we need to carry out before we can say that the effectiveness of an intervention model has been clearly established? How is our confidence in the effectiveness of a model affected by redesigns necessitated by applications to new problems and populations? Such questions may or may not be resolved through general principles, but they are certainly worth pursuing with that goal in mind.

This pursuit will be enhanced if the D&D paradigm itself can be exposed to the same developmental processes as the interventions it involves. The paradigm needs to incorporate a wider range of strategies to accommodate diverse developmental purposes and conditions.

CHAPTER SIX

Computer-Assisted Social Work Practice: The Promise of Technology

STUART KIRK, WILLIAM REID, AND CARRIE PETRUCCI

Yesterday morning, one of the authors used a laptop computer to retrieve and answer electronic mail sent from around the world, check the local weather forecast, and connect to the Internet to order books; then he drove to a bank in a car whose engine was partially operated by a computer to get cash from an automatic teller machine (ATM). Later, he went with a friend for a cup of coffee in a car that sported a computer monitor, accompanied by a synthesized voice that informed them when and where to turn to get to a restaurant.

Virtually everything we do at the dawn of the century is facilitated, monitored, recorded, or influenced by computers. Yet, the thirty-volume *Encyclopedia Americana* (1953) of the author's childhood, published halfway through the twentieth century, does not contain a single entry for "computers." In fact, a reference under the term "computers" sends the reader to an eight-page entry on "Calculating Machines" containing a description of the evolution of "adding machines" and the latest developments using punch cards to count and sort at superhuman rates. At mid-century, the technological developments that within four decades would transform practically every aspect of life in the United States—from manufacturing to commerce to personal communication—were unknown, except in the fantasies of science fiction writers. Every field of endeavor has been touched by this revolution and has had to grapple with how to harness technological promise for professional purposes.

The potential uses of computers and other innovations in social work were seen primarily by academics who thought that the new technology could increase agency accountability, make service delivery more rational, and improve practice. Administrators also foresaw possible gains in the management of resources and in record keeping. And, since the early computers were primarily rapid calculating machines for scientists and engineers, social work researchers believed that the new technology could advance the scientific basis of social work practice by offering a means of forging links between the research and practice worlds. But the advent of electronic computers made some social workers fear technologically driven change, and from the beginning they were ambivalent, suspicious, and fearful of creeping impersonalization or of machines replacing people.

This chapter explores how computer technology was introduced into professional social work. We will describe how rapidly the technology developed and how it was envisioned to apply to social work practice (see also Butterfield 1998). Not surprisingly, the development of technology far outstripped the profession's ability to fashion usable applications for it.

The Emergence of Computer Technology

The first digital electronic computer, the ENIAC, developed in 1946, helped the military to compute ballistics tables (Peuto 1997). This early computer filled a room, weighed 16,000 pounds, and cost $159,000 to $250,000 (Runyan 1991). Data were input through prepunched cards, an operation that itself required a fairly large and heavy machine. Because of the expense of operating mainframes, they were run in batch mode—an operator prepared the program and data offline, then ran the job all at once (Press 1993). The major barriers to the use of computers were their cost and the shortage of trained programmers. Throughout the 1940s and 1950s, computer use was limited to the government for handling defense and census data, and to large companies for performing large, repetitive data-processing tasks such as compiling actuarial tables. Few businesses and no individuals could afford computers (Runyan 1991). Social workers had little or no involvement with this technology; given the war-related origins of the new machines, reluctance to embrace them was understandable.

In the 1960s, broader use of computers was made possible by such developments as time-sharing; online applications; keyboards; and removable disk drives (Peuto 1997; Runyan 1991). Progress at leading universities raised the possibility of computers being available to an even wider audience. For example, innovators at Stanford Research Institute invented word-processing applications, the mouse, the multiwindow display, automatic word wrap, and functions for deleting and searching text (Press 1993). Furthermore, software was developed with interactive use in mind. Computers moved from accounting and statistical computations to real-time transaction and information system processing (Press 1993; Runyan 1991).

By the end of the 1960s, large businesses and government agencies everywhere realized that computers could help with their data processing. Still, small businesses and social work agencies had yet to see any immediate benefits from the technology. Computers were invisible machines in distant locations. Only the increasingly common IBM punch cards provided tangible evidence of them. Computers remained very large; they still took up an entire room and were incredibly expensive to run. Monitors and keyboards were only beginning to be utilized by a few specialists. That a high-powered computer might be small enough to sit on someone's desk could only be imagined by a few (Press 1993).

Nevertheless, the promise of computers was widely recognized. Given their military origins and the fact that for many, they symbolized an increasingly bureaucratic, impersonal, and inflexible form of social organization, it was to be expected that their use in other arenas would be controversial. And so it was in social work. In fact, the first two articles about computerization to appear in the social work literature reflect radically opposing viewpoints. The first warns of the dehumanizing effect of computers, while the second promotes the use of computers for dual data-processing and research purposes. These articles frame the promise and pitfalls of computer-assisted social work practice.

Opposing Views of Computers

The earliest article we could find about computers in the major social work journals appeared in *Child Welfare* in 1967 and was titled, "Social Casework—Science or Art?" (Teicher 1967). In it, Morton Teicher, a dean of a school of social work, rails, in what reads like a rousing speech to an antitechnology group, against the notion that social casework can be a form of science and that computers have any useful role in practice. He stresses that casework calls for the creative skill of the artist and warns about the dehumanizing effects of computers that will eliminate individual worth, human freedom, and responsibility. He describes game theory as a recent application of computers in the military and business and argues that the use of quantitative data in social work is inappropriate because social workers deal in values that are not reducible to the mathematical language of computers. These machines, he argues, require clear and structured information that is simply not available for most social work situations. Too many idiosyncrasies about the client and the situation are unknown. People are too complex to convert to numbers for processing, and to do so is to ignore individual qualities. Teicher identifies a conflict between intuition and intellect: practice is doing and feeling, not scientific investigation. Social work offers skill that is based on knowledge and experience, used with conditional intuition, refined hunches, disciplined instinct, and tempered faith. Computers are to Teicher a threat to the profession.

A year later in *Social Casework*, in the second article about computers to appear in the social work literature, "Developing a Data Storage and Retrieval System" (1968), Ivan Vasey, a physician and director of a mental health agency, waxes enthusiastic about how computerization can benefit social agencies. His article is a case study of how "modern" electronic data-processing techniques can improve the efficiency and quality of social services at a clinic without increasing the staff workload. Today his description seems elementary: developing common forms, gathering client data, having the data key-punched on cards, involving staff in these developments, and producing simple reports. The clinic used a time-share computer at another location. His agency entered more than 1,200 cases with up to 400 items per case into a database; it also entered information on 100 control families and then analyzed follow-up data on its clients, thus merging administrative and research functions within a direct practice agency.

Vasey's article traces the process of integrating a computerized data-processing system into social work practice and foreshadows what will become core elements for success: involve all levels of staff from the beginning so that the system is useful and they have a stake in it; coordinate face-to-face communication between data-processing staff and program staff during development of the system, not only to decrease the possibility of animosity but also to develop a system that is meaningful to program staff and feasible for the computer to run. He also notes the positive "unintended consequence" of the staff becoming interested in the data produced by the system and pursuing research questions as a result, an embryonic version of the practitioner-as-researcher concept pursued in the following decade. He concludes that a computerized record-keeping system can improve services and save staff time, that clerks are capable of extracting and coding data reliably, and that his staff showed great interest and excitement in being a part of a "forward-looking" agency.

Clearly, Vasey and Teicher were viewing different possible roles for computers, one a dream, the other a nightmare. The two articles appeared at about the same time, but neither author appeared to be aware of the other's viewpoint; in fact, Teicher's article has no citations and Vasey's has only two. Vasey was enthusiastic and optimistic about the role of computers in modern agencies, while Teicher was fearful and pessimistic about how they might completely distort and dehumanize practice. Vasey concretely describes practical applications of computers that appear to avoid the pitfalls that Teicher, whose concerns are more abstract, warned about. For example, Vasey uses the data-processing system primarily for documentation and retrieval of straightforward information, such as demographics and characteristics of case plans, not for decision making, which Teicher worries about. Teicher's denunciation of computers in social work seems to eliminate even the possibility of using them in the noncontroversial ways that Vasey illustrates.

These two early articles, however, offer a glimpse of developments to come. Vasey foreshadows the optimism that computers could serve in the quest for efficiency, the integration of research into practice, improved services, and communication. Teicher's perspective anticipates harm to clients and workers and dehumanization of social services. We will use their hopes and fears to analyze the ways that computers have been used in social work practice. Before turning to that task, however, we need to provide a brief history of the evolution of computer technology.

From Mainframes to Minis to PCs

From the late 1960s to the late 1980s, computers moved from mainframes housed only at university computer centers to the tops of small desks everywhere. Microprocessors, the chips that control the speed of computers, effectively eliminated the need for room-size machines. The minicomputer, a smaller version of the mainframe, brought computers closer to the user (Runyan 1991) and decentralized computer use by 1981 (Strassman 1997). These smaller computers could run both word-processing and data-processing programs, making them a staple in the office environment (Runyan 1991).

Microcomputers further accelerated this trend. Individuals owning their own computers, rather than sharing a mini or mainframe, was an enormous technological and cultural breakthrough (Press 1993). The first microcomputer sold to the general public in the late 1970s was the Altair (Miller 1981), which appeared on the cover of the magazine *Popular Electronics* and took the personal computer industry by storm. Still, it was no ready-to-use gadget; it came as a kit for $397 and required hand-soldering. To make the Altair usable, one had to purchase a monitor, floppy disks, and a printer. This hiked the price up into the $5,000 range, about the cost of a small car, but for the first time it was possible to have a complete computer system on your desk that could run programs written in Basic or Fortran (Press 1993). Hobbyists and entrepreneurs bought the Altair by the thousands (Dvorak 1998). Computers had now truly become available to "the masses," although the early consumers were computer geeks with soldering tools.

With the development of microprocessors, word processing, and spreadsheet software programs, the demand for personal computers ballooned among small businesses (Peuto 1997; Runyan 1991). Although mainframes and minicomputers were still the most reliable and effective means of large-scale data processing at the end of the 1970s, personal computers were poised to dominate the next decade (Spencer 1999). For example, by the end of the 1980s there were 50 million PCs in the United States alone and more than 100 million worldwide. Sales amounted to a staggering $37 billion just for personal computers, plus an additional $56 billion for peripheral items (Spencer 1999). Many people who had never used any type of computer in 1980 were unable to function without one both at work and at home by 1990. The meteoric rise in the status of PCs was dramatically illustrated at a celebration in Philadelphia marking the famous mainframe, ENIAC's, thirty-fifth birthday. The ENIAC was pitted against a Radio Shack TRS-80 microcomputer in a race to square all integers from 1 to 10,000. The mighty ENIAC did it in six seconds. The tiny TRS-80 accomplished it in a third of one second (Spencer 1999:106).

There was an explosion in software development. For example, by 1982, more than 17,000 software packages were available for use with Apple II computers. Software gadgets, such as the familiar Macintosh icons of the clock, the smiling computer, and the trash can, were developed to create a user-friendly environment. Relational databases, graphics programs, and touch screens all added to the ease of use. Sales in the software industry reached $1 billion by 1983, and grew at the phenomenal rate of 50 percent each year (Spencer 1999). In 1985 Microsoft marketed its Windows operating system. Other technology was also advancing rapidly. Innovations included a mouse-supported operating system, hard disk drives with a larger capacity for memory, sound cards, 3.5" disk drives, and color printers (Spencer 1999).

New uses for computers were sprouting everywhere, in the worlds of finance, manufacturing, medicine, and education. Computer technology was also becoming unavoidable in everyday life, in such forms as the scanning technology at checkout counters and the 72,000 ATMs throughout the country by 1988. As it became more and more affordable, even for nonprofit agencies and individuals, computer advocates wanted social workers to think about how to use this technology in service delivery.

The Quest for Efficiency

Computers have a great capacity to store information in microscopic places and then retrieve, manipulate, and display it quickly in various forms. The possibility of replacing written files held in manila folders in rooms full of filing cabinets with small electronic records was revolutionary. Social work agencies, like other organizations, saw data storage and record keeping as the primary uses of computers into the early 1980s (Geiss 1983). Every agency had a need to know how many people were served and who they were, and where the money went. With written case records, answering such simple questions was a labor-intensive task, but with data-processing systems, information such as a client's address, age, or gender could be accessed quickly and effortlessly. Keeping track of expenditures was also an easily computerized task because the information is quantitative and highly structured. A *Practice Digest* article for direct practitioners heralded the personal computer as an efficient means to maintain a billing and accounting system (*Practice Digest* 1983), among other uses. One practitioner (Clark 1988) pointed out that his computer was his secretary, accountant, office manager, and research assistant. The ability to keep track of masses of information efficiently, however, was quickly seen as a means to other ends.

ACCOUNTABILITY

Mitchell I. Ginsberg, a prominent social work leader, warned in 1968 at a Council on Social Work Education annual meeting that social workers would be held "publicly accountable" for program effectiveness, given the massive spending in welfare programs (Fuller 1970). Calls for accountability became a professional mantra during this era and served as a backdrop to many of the developments discussed in other chapters of this book. Since accountability required keeping track of mountains of information about clients and services, it is not surprising that many would think that computers could help (Boyd et al. 1978; Fuller 1970; Hoshino and McDonald 1975; Reid 1974). The quest for efficiency in data management became a potential means of studying program effectiveness.

George Hoshino and Thomas McDonald worked this new theme into an article, "Agencies in the Computer Age" (1975), arguing that the computer can be a "vital tool" in allowing ordinary agencies to increase their accountability. Computers can be used for ongoing program evaluation to measure performance and efficiency. One part of the article is a "how to" description of this new technology and its operation.

By the end of the decade, Lawrence Boyd et al. (1978) found 31 articles about computers in the social work literature. Accountability was among the most common themes, but, they note, in reality the most common use of computers was for clerical and record-keeping duties. While this might be indirectly relevant to accountability, the more sophisticated uses of computer technology, such as program evaluation and information systems, were at a primitive stage of development. Furthermore, Boyd et al. found little evidence that service providers benefited directly from the new technology or were using it to carry out their professional tasks. The capacities of computers to handle data continued to expand, and soon record keeping was renamed management information systems.

MANAGEMENT INFORMATION SYSTEMS

With success in data processing, staff and management became more willing to take on more complex tasks with the computer, such as using management information systems (MIS). One of the immediate benefits of MIS is the ease with which reports can be generated, saving considerable staff time, tedium, and effort (Schoech 1995). Designing and implementing MIS, however, was no easy matter. Like any innovation, it took time and sensitivity to develop and introduce into an agency (Schoech et al. 1982; Wodarski 1987; Pardeck 1998). Typical management information systems are either "client" systems that provide similar data for many clients or "staff" systems that provide standardized information for all staff. One of the earliest social work examples of a client system was *Childata*, developed in the late 1960s by a consortium of child welfare agencies in Chicago (Rothchild and Bedger 1974). Data about children under care, including type of care arrangement, services needed and planned, adoption prospects, and goal achievement ratings were submitted by each agency to a time-sharing computer system, which would then send reports to all the agencies. Special reports would be automatically produced on children who had remained in certain statuses, such as temporary foster care, beyond designated limits.

Not everyone was persuaded of the importance of MIS. Ram Cnaan (1989) provided six reasons why social workers should adopt and integrate information technology. First, the new technology entailed the making of new rules, and social workers should be involved in this process. Second, computer technology had already become a permanent fixture in society, and social workers should be involved in developing it so that professional values could be integrated into emerging systems. Third, social workers would suffer a political loss if they missed these important developmental phases, resulting in systems that would not contain their professional values. Fourth, the use of information technology provided an opportunity for social work to clarify and strengthen its knowledge base, which could translate into more effective service for clients. Fifth, records are easily retrievable when data are recorded and monitored with a computer system. Finally, the new technology can allow for knowledge dissemination among practitioners, so that information can be shared and issues debated (Cnaan 1989). MIS, Cnaan concluded, should be shaped by social work practitioners to improve services and increase knowledge development.

However, the primary focus of MIS was efficiency. A report of the success of an information system for a county general relief program by Velasquez (1992) is representative. An MIS reduced the time it took to process applications, increased productivity, increased intakes scheduled per month, recouped the system costs in fourteen months, reduced processing errors, and improved staff turnover and morale (Velasquez 1992).

In general, the most common applications of MIS were for administrative monitoring and accountability, such as in quality assurance programs (see Auslander and Cohen 1992). Practitioners were less likely to use MIS, although these systems could be designed to assist with clinical tasks such as recording assessments, treatment plans, and client contacts and keeping track of data related to implementation and effectiveness (Wodarski 1987). In fact, some "clinical information systems" were developed specifically for use by practitioners. Benbenishty (1989) encouraged individual workers within agencies to collect information that would then be aggregated at the agency level. He believed that practitioners can learn more about their own clients by comparing them to all the clients in the agency. To be effective, the clinical information system must be integrated into everyday practice by practitioners themselves. One-shot or intermittent research projects would not accomplish this. He made many suggestions, among them that the clinical information system be composed of single case studies submitted by all practitioners and that data on current as well as past clients remain in the system so that ongoing practice can be monitored through comparison with past data. (See also chapter 4.)

Supporters assumed that computer technology was a *must* in social work and that the profession was embarrassingly behind others in using it. Cnaan (1989) offers several explanations for this lag, related to inherent qualities of social work and to practical and political realities. Among the inherent barriers are that social work knowledge is not easily quantifiable and that the culture of social work simply clashes with the culture of computers. The practical and political barriers to integration include lack of available resources to purchase, develop, and maintain information systems. Furthermore, because each agency essentially requires its own system, there is a very small market and therefore little incentive for software developers to design agency-specific information systems. And since software development is not usually considered as scholarship in tenure decisions, the academicians who might engage in it are not encouraged to do so. Finally, social workers have a common fear of "overbureaucratization," and the systematic, standardized structure of information technology suggests this as an unintended possibility (Cnaan 1989).

Nevertheless, there were early attempts to use MIS to support practice. For example, Friedman (1980) presented an early application of an information system. In his study, sixteen behaviors were operationally defined and monitored in a youth population in a residential treatment center. The center used a token economy treatment program in which patients carried computer cards that were marked by staff. The data from the cards were then entered into SPSS and analyzed using a remote computer. Daily reports were produced tracking youths' behavior and comparing it to the entire cohort. The goal was to build an information database so that treatment decisions could be made based on data from a cohort of youths rather than on opinions of senior staff, proximity to the situation, or the heat of the moment. Such early uses of management information systems to assist clinical decision making would be developed later into decision-support and expert systems.

By the 1990s, management information systems had become commonplace. The federal government, for example, required child welfare agencies to participate in one, the Statewide Automated Child Welfare System, or SACWIS (Department of Health and Human Services 2000). Individual agencies also developed their own systems, for example, the one at Boysville in Michigan that integrates clinical, supervisory, and administrative information (Grasso and Epstein 1989) or the one at the Danville VA Medical Center (Breeding, Grishman, and Moreland 1996).

The Quest to Integrate Research Into Practice

Many commentators saw computers as a means of promoting research within agencies and allowing practitioners to do research. Fuller (1970) was the most optimistic. He believed that the two greatest barriers to research in practice settings were lack of time and lack of statistical competence. He argued that computer technology brought "research within the range of almost every practicing social worker" (608) and saw no reason why every practitioner couldn't learn the elements of Fortran programming in order to do statistical computations. He thought that with a little computer training and some familiarity with Monte Carlo techniques, practitioners would become applied researchers.

He was not alone in the quest to integrate the research and practice worlds via computers. Hoshino and McDonald (1975) made a similar claim, suggesting that with only a few hours' training, agency staff could be taught to use early versions of SPSS, a statistical package for analyzing data. Indeed, computer technology and the statistical packages put data analysis capabilities in the hands of nonstatisticians for the first time. But was it realistic to expect that nonresearchers, particularly the math-phobic types who gravitated to careers in the helping professions, would embrace computer technology, Fortran programming, or sophisticated statistical software? Hesitancy, resistance, or neglect could be expected.

The very nature of applied research itself was thought to be changing in agencies. For example, Reid (1974, 1975, 1978) foresaw the agency functioning as a "research machine," which would use computerized information systems to generate a limitless variety of specific studies at a fraction of what they would cost if carried out by traditional methods. Some spectacular examples of this kind of research did appear. For instance, Fanshel (1977) used a multiagency information system in New York City to produce a study of parental visiting of children in foster care; the sample consisted of more than 20,000 cases. However, this mode of research did not become the major force that Reid and other computer advocates had predicted. The notion that sophisticated information systems able to generate research studies would rapidly develop had turned out to be excessively optimistic.

Computers could support many research functions, but agencies would need specialized personnel to analyze their data quickly in order to provide a feedback loop useful to practitioners (Reid 1974). Hoshino and McDonald (1975) promoted this research-in-the-practice-setting model, thinking that there would be no additional dollar costs if, in the name of accountability, agencies evaluated their own service delivery on an ongoing basis. In their case study, they described how an MSW student used a computer to study forty-eight case files. The project demonstrated the feasibility of this approach, but it terminated when there was no agency staff person to take over the research role after the MSW student departed. They recommend that to avoid this, research must be integrated into everyday agency practice; staff involvement is crucial in identifying important outcome variables and interpreting findings (Hoshino and McDonald 1975). This advice, first offered by Vasey (1968), would be restated again by many authors in the coming decades.

By 1990, with greater practitioner and agency access to computers, there was great optimism that computers would transform practice routines. For example, a report by Merlin Taber and Louis DiBello (1990) describes a demonstration project that trained staff from twelve agencies to use computers at a time when some of these agencies did not have even a single computer onsite. The project's objective was to develop and use a computer program to input data about everyday worker tasks, including information on clients, services provided, and outcomes. Taber and DiBello trained front-line staff because they believe that those who actually use the computers can undermine or enhance the implementation of an information system. The authors valued staff feedback, which kept them alert to how overloaded personnel could be and how computers could ease this workload (Taber and DiBello 1990). They placed the needs of the practitioners before the requirements of the computer program, and attempted to make the program relevant to staff and maintain low start-up costs. Three agencies actually implemented the database system. Several more put in requests for computers. And in a pre- and post-test evaluation, most participants were much more positive about the use of computers as the project ended than when it began. But computers were not yet being used to examine caseloads or to improve practice (Taber and DiBello 1990).

SINGLE-CASE RESEARCH

Clinical researchers interested in promoting the use of single-subject designs (see chapter 4) to make practice more scientific quickly saw how computers could be employed to inform practice, increase effectiveness, and promote accountability (Benbenishty and Ben-Zaken 1988; Bronson and Blythe 1987). Computers would make data analysis faster and easier, thereby circumventing the common barriers of practitioners' lack of statistical know-how, time to input and analyze the data, and agency support (Bronson and Blythe 1987).

Denise Bronson and Betty Blythe (1987) made an early attempt to do this through the Computer-Assisted Practice Evaluation Program, or CAPE. CAPE was developed in a spreadsheet format using Lotus 1-2-3, which supported the entry and analysis of single-case AB design research. The social worker would enter data, and the program would generate graphs and statistical tests in a minute. But it was not quite that easy. In fact, the use of CAPE required a basic knowledge of statistical procedures and interpretation, often lacking among practitioners. Although CAPE was useful for some, it clearly had a long way to go to be useful across different types of single-case research designs and circumstances (Bronson and Blythe 1987).

In another demonstration project using single-case evaluations, a computer system for the Young Families Project allowed practitioners to enter client data and generate graphs and statistical analyses. The agency utilized a task-centered approach, with an emphasis on behavioral change. Benbenishty and Ben-Zaken (1988) report that social workers using the computer technology had difficulty performing certain standard tasks required of the program, such as assigning numeric values to certain client characteristics. Achieving the consistency and uniformity needed in entering data on every client was difficult, but once the initial assessments were performed and entered, social workers did not find that analyzing the data was too time consuming or stressful. The authors note, however, that the technology was not fully integrated into the agency's daily operations and that researchers may be too sanguine about the benefits of the new technology (Benbenishty and Ben-Zaken 1988). These and other problems in using single-subject designs are described in chapter 4.

The Quest for Improved Service

There is a striking contrast between the fast-paced, exciting, rapid evolution of computer technology and the cautious, conflicted attempts of social workers to adapt this technology to practice. With the advances in microprocessors and software, there was a growing recognition that computers would eventually be available at all levels of practice and that direct practitioners and smaller, nonprofit agencies would no longer be left out. The incredible advances in hardware and software seemed to make the dream of agency-based research a realistic possibility. But besides the long-term goal of building an objective knowledge base, there was more immediate interest in using computers to enhance the quality of services to clients. This interest was expressed in a variety of ways.

EARLY VIEWS

Relatively early in the evolution of computer technology, a few social workers could envision an expansive role for it in practice. For example, Paul Abels offers an upbeat and hopeful view in "Can Computers Do Social Work?" (1972), noting that social workers were well behind psychiatrists and psychologists in using computers for tasks such as interviewing and assessment. He characterizes resistance as due to the "antihuman" perception that social workers have of computers. Fred Vondracek and his colleagues (1974) argue that computers can greatly improve service delivery by promoting better coordination and integration across agencies. Computers are tools that liberate direct service practitioners from menial tasks so they have more time to address humanitarian concerns. Technology can enable and enhance the quality and quantity of services to clients by removing the mounds of paperwork and red tape endemic to bureaucracies. Boyd et al. (1978) also stressed the support that computers could provide to direct service practitioners, which was essentially unrecognized because social workers did not understand it well.

These authors represented a small but persistent group of staunch promoters who saw the potential of computer technology for assisting with important direct practice tasks such as intake, assessment, record keeping, data analysis, decision making, and coordination of services (Abels 1972; Fuller 1970; Hoshino 1975; Schoech and Arangio 1979; Vondracek 1974). Fuller (1970), for example, recognized that computers could assist in the intake procedure by automating the application process through remote access and writing social histories based on data entered by clients or staff. Similarly, a "natural language intake summary" could be generated by the computer based on data entered in client intake forms (Vondracek 1974). Abels (1972) went a step further in predicting that computers could actually be used to conduct interviews or make assessments, keep records, and make diagnoses. He noted that computers can alter the line of questioning, are more objective than humans, and do not become harassed when over-worked. Schoech and Arangio, in a prescient article, "Computers in the Human Services" (1979), published in NASW's major journal, *Social Work*, were correct in many of their predictions, including that computers would become standard tools in social welfare agencies, whether practitioners were ready for them or not, and in typical middle-class homes by the early 1990s. These authors suggested that computers would soon be capable of setting workers' schedules through beepers, planning their routes, and monitoring their work (Schoech and Arangio 1979). This scenario had a "Big Brother" ring to it, which was unlikely to endear it to front-line work-ers.

Many authors foresaw more moderate and immediate benefits. Fuller (1970) stated that computer technology could increase the "scope of case-work" by increasing the number of clients who could be helped, in part through faster intake procedures. George Hoshino and Thomas McDonald (1975) noted that computer data analysis would allow screening procedures to link clients to services. Automating the intake procedure would avoid duplication of work and allow quick retrieval of already existing client files. In addition, client data would be gathered more uniformly and completely (Vondracek 1974).

Computers could make service integration across agencies—a promi-nent concern in the 1970s—a reality (Schoech and Arangio 1979), thereby delivering services more economically to more clients (Fuller 1970). Service networks could match agencies to client needs based on a common intake form, and community agencies could work cooperatively and economi-cally. Vondracek (1974) reminds social workers that this type of computer-assisted coordination of services could help fulfill one of the profession's oldest goals: the desire of the original Charity Organization Societies to make service delivery rational.

As much as management or clinical information systems would repre-sent a major advance over the traditional hand-written or typed client case records, they were only the beginning of what might be achieved by using computers in social work practice. Before the full implications or implementation of MIS in most agencies had been realized, suggestions were made that computer technology could be harnessed for the most sacred of professional activities, clinical decision making.

DECISION SUPPORT SYSTEMS

Decision support systems are a natural outgrowth of data processing and MIS (Schoech 1995). A decision support system is a computer-based processing application designed to help professionals make complex decisions (Schoech and Schkade 1980), such as those prompted by the question: What if this were the case? The systems answer these questions by using statistical models based on algorithms. Schoech and Schkade (1980) were among the first to advocate using decision support systems in social work, years before practitioners were even considering, much less accepting, them and at a time when even social work scholars were barely grasping the notion of inputting client records for simple retrieval purposes. Schoech and Schkade saw computers as a form of assistance rather than replacement, and argued that decision support systems in child welfare were needed because high staff turnover resulted in many inexperienced workers and poor-quality case records. They suggested that the function of such a system was to be a "friendly helper" to enhance the decision making of social workers, not replace them. They thought that decision support systems could be the most valuable application of computer technology in human service agencies.

At the time, this was a bold prediction, given that personal computers were not readily available and mainframes or offsite computers were hardly user friendly. Boyd et al. (1981) developed and studied the integration of a decision support system in a large in-home supportive services agency, with the goal of regulating equitable allocation of awards among clients across three county offices. These studies point out various ways that a decision support system can be helpful to caseworkers: assisting with routine tasks such as filling out and printing forms; helping a worker through the decision steps; offering options for unstructured queries; and generating reports that explain decision making.

Implicit in both of these case studies was the use of decision support systems in monitoring staff and accountability. In a field where independent decision making had been the norm and computer technology was foreign, some practitioner resistance to decision support systems could be expected, especially if the technology was not introduced properly. Later studies of decision support systems for child welfare placement decisions (Schwab et al. 1985, 1986) and for in-home support services resource allocation (Rimer 1986) were successful, however, in part because they focused on discrete decisions and allowed staff to help develop the system (Schwab et al. 1986).

Decision support systems were also expected to play a role in organizing and disseminating the profession's knowledge base. Some believed that they could help by translating the knowledge of experienced workers so that it could be utilized by others (Schwab et al. 1986). For example, Schwab et al. explain that expertise in child welfare cannot be easily articulated, captured, stored, and passed on to future generations of child placement practitioners. To remedy this, the authors developed a software program that could make placement recommendations based on a database containing the records of 2,799 children in various types of placements.

EXPERT SYSTEMS

Professionals are proud of their expertise—the ability to use their knowledge and judgment together to diagnose and solve problems—and consider it a defining characteristic. Expert systems are more advanced than decision support systems because they attempt to model or simulate professional decision-making processes themselves. The idea of a computerized "expert system" stimulated practitioners' two greatest anxieties about computers: that they would dehumanize the practitioner-client relationship and that they would eventually replace practitioners. Neither of these scenarios has to come to pass, but the great optimism of the 1980s and the untested capacities of computers made anything seem possible. After all, technology was replacing workers in other fields, such as manufacturing, and it did not seem outlandish to extrapolate a similar pattern in social work and elsewhere. This line of reasoning was encouraged inadvertently by advocates of expert systems, who suggested that they could be used by "untrained" or "less experienced" workers (Schuerman 1987). This implied that experienced workers could be replaced by paraprofessionals relying on well-developed computer programs to make important decisions in fields such as child welfare or mental health.

Expert systems grew out of the larger field of artificial intelligence, in which computer-generated behavior imitates how the human mind works (Butterfield 1987). Expert systems were identified as the most promising form of artificial intelligence, and for a brief period there was great enthusiasm among some scholars about their potential role in social work. Promoters of expert systems expected that they could be used to pass on scarce expertise to less experienced workers (Gingerich 1995; Schuerman 1987); to train practitioners (Gingerich 1995; Goodman et al. 1989; Schuerman 1987); to "guide and monitor" practice through documentation of the implementation of an intervention; to acquire, describe, and refine practice knowledge for later use; and to develop theory by carefully outlining the concepts used to solve problems (Goodman et al. 1989; Gingerich 1995). In short, expert systems were viewed as ways to identify the profession's knowledge base, rationalize decision making, and capture for both science and the profession that elusive ghost, practice wisdom.

Three components make up an expert system. First is the knowledge base (Schoech et al. 1985) or set of rules (Schuerman 1987). Second is an inference mechanism (Shoech et al. 1985; Schuerman 1987). Third is the set of facts related to a particular case that are known and input (Schoech et al. 1985; Schuerman 1987). Uncertainty is accounted for by attaching levels of statistical probability or uncertainty to facts or rules. Verbal levels of uncertainty can be programmed in lieu of numerical probabilities (Schuerman 1987).

Development of an expert system is driven more by the need to document knowledge than by the development of a computer program. It requires documenting the knowledge base of social work practice (Gingerich 1995; Goodman et al. 1989), consulting with seasoned practitioners who are "domain experts" and with computer programmers (Carlson 1989), gaining formal validation by comparing the results of the expert system with the judgments of domain experts, and using computer programs or "shells" to construct the expert system (Brent 1988; Gingerich 1995).

These are often not easy steps to accomplish. Schuerman et al. (1988), for example, in developing an expert system in child welfare, identified several intellectual issues that must be considered, including the integration of "common sense" into the knowledge base; consideration of the context of the organization, such as political barriers, caseloads, and resources; dealing with what is actually done in practice versus what may be considered "best practice"; and acknowledging that social work practice is "moderately complicated," which makes integration of an expert system challenging (Schuerman et al. 1988). For example, care must be taken in gathering the knowledge from the expert staff, and validating the expert system is difficult, due in part to conflicting expert opinions about what constitutes a "correct" decision (Schuerman 1988).

Matthew Stagner (1994) is one of the few in social work to describe in detail the process of developing an expert system (for a child welfare agency over a three-year period), a process similar to the development of "grounded theory" (Glaser and Strauss 1967), in which the researcher creates a model based on observations in the field. Stagner carefully outlines each step in the development process. The research team was made up of a "domain expert" (child welfare experts), a "knowledge engineer" (the researcher), and a multidisciplinary review team. Perhaps of greatest value is Stagner's description of the actual process of knowledge acquisition from the experts. Five were selected from those suggested by colleagues and superiors. They were also chosen for their ability to conceptualize and articulate their decision-making process. Workers were selected who could attribute their decision making to a clear set of observations and facts rather than to intuition or gut instinct (Stagner 1994). They had to be willing to have their decisions closely scrutinized, be available for several meetings over the course of the study, and have a tolerance for the detail work necessary. A close relationship between the expert and the researcher had to be developed.

In the end, the experts and researchers were satisfied overall with the process, although the project had to overcome various practical and conceptual difficulties in the development of this prototype system. It is unclear whether the system was later implemented. Stagner concludes that as long as experts can articulate their decision-making process and that process can be validated by the review of additional experts, knowledge acquisition can occur and be translated into computerized versions of expert systems (Stagner 1994).

Jaffe (1979), who also developed an expert system of child placement decision making, is aware of concerns about the potential for dehumanization but makes a case for using technology for the rational provision of services. He recognizes that the values and principles underlying clinical practice are important but argues that good clinical practice should be clearly communicated so that it can be replicated and verified. He recommends that practice be conceptualized and operationalized in ways that make it accessible to study. He foresees the synthesis of art and technology in the next decade and hopes this will reduce the tension between researchers and practitioners.

Despite initial enthusiasm among a few advocates and fears among social work practitioners, expert systems in social work have not developed very quickly and are not used in routine agency practice. The early promise that they would allow researchers to uncover professional knowledge and expertise, codify it, and make it available for dissemination has not been fulfilled. In the growing body of research, it is a rare article that reports the development or testing of such systems in practice. Expert systems can only be as good as the knowledge base on which they are built. In social work, the knowledge domains are difficult to systematize, practice experts often do not agree, and many times, not all facts are known (for discussion, see Mullen and Schuerman 1990; Stein 1990; Reamer 1990; Wakefield 1990). If expertise has not been developed, then it cannot be translated into an expert system. Additionally, expert systems hinge on the necessity of making discrete decisions, e.g., whether to remove a child from a home. But often practice is not driven by the need for a specific decision, as when a worker is simply gathering information, making an assessment, and establishing a relationship. Although computerized expert systems never made much progress in the practice arena, computers made more headway in assisting with assessment tasks.

CLINICAL ASSESSMENT

The nature of some practice tasks makes it easier to capitalize on computer technology to facilitate them. Initial assessment of clients is one. Although the use of computerized information systems, expert systems, and decision support systems has been limited, computerized assessment has been more easily accepted. Perhaps this is because the role of this technology is clearly to assist practitioners rather than to replace them, or because its time-saving capacity is clear. Assessment involves the collection of various small, discrete pieces of information through self-reporting, usually following a structured or semistructured format compatible with computer data entry methods.

The undisputed pioneers in computer-assisted assessment in social work have been Walter Hudson and his associates. Through the development of the Client Assessment System, or CAS, they have integrated current computer technology with up-to-date, empirically tested assessment tools. CAS was created with the computer novice in mind and can be used interactively with the client and/or the social worker. It contains twenty standardized multiple-item assessment scales, such as a global screening inventory, partner-abuse checklists, and behavioral checklists for children, that have been developed and validated over several decades. Clients can input the intake data, or social workers can do so in an interview format (Nurius and Hudson 1988a, 1988b). The CAS has a very user-friendly interface, so almost anyone who can reach the keyboard and read the screen can perform the assessment. As important, CAS is structured in such a way that agency-specific assessments can be added. Finally, CAS facilitates client monitoring and evaluation and accommodates the many functions necessary throughout the life of a case, such as transferring, writing progress notes about, or closing it. Nurius and Hudson (1988a) are quick to point out that a computer cannot replace clinicians. Computers cannot do all that a practitioner can do, such as noting the demeanor of a client or easily recognizing important idiosyncratic contextual factors. In short, CAS was developed by social workers for social workers, and while it has its limitations, its benefits are evident (Franklin et al. 1993).

Computer-Assisted Interviews. Social science research shares with social work practice a heavy dependence on respondent/client interviews for relevant information. In survey research the standard method of data collection was in-person interviews in which the social worker would use a structured and at times very complex written questionnaire to elicit responses from an interviewee. For reasons of cost, many survey and marketing organizations eventually opted for telephone interviews. Then, with the development of computer technology, they began to use computer-assisted telephone interviews (CATI), in which the interviewer follows the prompts on a computer screen in asking questions and recording answers, thereby eliminating the need for a separate data entry step (Baker 1992). An additional innovation has been computer-assisted personal interviews (CAPI), in which the interviewer uses a small laptop computer in conducting face-to-face interviews. Thus, the programmed computer plays a direct role in guiding the interaction between interviewer and respondent, setting the pace of the interview, checking the responses and response patterns immediately, and jumping through the various contingency sections of the questionnaire. And, unlike in computer-assisted telephone interviewing, the computer's presence and role are openly visible to the respondent.

The goal of CAPI is to make interviews faster, more accurate, and cheaper, and make complex questionnaires easier to implement (Baker 1992). At first, practitioners worried about the impact the computer might have on the quality of the interview. But fears about confidentiality, the loss of eye contact between the respondent and the interviewer, and the intrusiveness of the computer have not been justified. In fact, most respondents find the experience of CAPI to be enjoyable and interesting (Baker 1992). Interviewers' original concerns that the software would be difficult to manipulate and that the quality of the data (particularly with open-ended responses) would suffer have proven unfounded. Interviewers have used the CAPI software on laptops in the field with minimal problems. Potential errors are eliminated because the computer alerts interviewers if they use an incorrect skip pattern, enter an inconsistent response, or miss a response. Also, no difference in data has been noted between the open-ended responses on CAPI and and those recorded with paper and pencil. There is no reason to believe that the use of CAPI has any adverse effects on the quality of the data, although there are still unexplored issues regarding the populations with whom such methods are acceptable (Baker 1992).

The use of CAPI has some obvious relevance for intake assessments, needs assessment, outcome and follow-up measurement, and studies of client satisfaction with services—traditional clinical tasks that use scientific survey methods. To date, however, CAPI is still primarily a tool of researchers in social work, rather than a routine method in practice. Nonetheless, this computer application could easily be adopted for routine client assessments and evaluation of service outcomes.

Virtual Reality. Computers creating virtual environments have been used in treatment. Gary Holden et al. (1999) used STARBRIGHT World, a three-dimensional, multiuser, virtual environment to link seriously ill, hospitalized children into an interactive, online community. The children participated in a cartoonlike world that appeared on their computer screens by selecting a character to represent them and moving that character through the virtual reality to encounter and communicate with other characters (hospitalized children) doing the same thing in real time, via voice and videoconferencing technology. The study attempted to determine whether use of such technology in pediatric care would decrease levels of pain intensity, pain aversiveness, and anxiety. The results were equivocal, but the intervention represented an application of computer technology to direct practice.

The Quest for Communication

The 1980s brought computer hardware and software from distant sites staffed by computer technicians into nearly every agency and onto the desks of many managers and practitioners. The presence of computers was no longer something to be imagined; it was tangible. As momentous as the technological advances of the 1980s were, few could foresee the transformations coming in the next decade. The rapidly increasing capacity of hardware and more sophisticated software paved the way for much broader uses of personal computers beyond storing, retrieving, and manipulating data. In the 1990s, computers were transformed from machines for number crunching and word processing into indispensable instruments of communication and information dissemination.

Currently, it is hard to imagine the working world without e-mail, instant access to the Internet, or Web pages. Yet, little more than ten years ago, the software technology that would allow these innovations to become household words had not been invented. But the Internet was not an instant success. In fact, not until 1994 was it featured on the cover of a computer magazine, and even then, some believed that the Internet was only for scientists and universities (*PC Computing* 1998).

Although this technological revolution is just in its infancy, it has rapidly altered the ways that many people work and communicate in many fields. How will it affect direct practice? Since social work has usually lagged in adopting technologies, it is probably too early to know, but already there is at least one handbook about the Internet for social workers (Grant and Grobman 1998). The literature contains only a few suggestions of what the future may hold, and information can become dated almost before a report is published. Here are samples of diverse current uses of computers for communication.

ONLINE SUPPORT GROUPS AND CONFERENCING

The new modes of communication have relatively quickly been pressed into service to extend the help provided by practitioners. It is only a short leap from e-mail and chat rooms to online support groups and therapy sessions (O'Neill 2001; NASW News 2001). Computers and telephones are being used as the media for self-help and support groups, the latter moderated by a therapist (Meier 1997; Finn and Lavitt 1994). An early survey of 213 group workers revealed only 6 had experience with telephone groups and 7 with computer groups (Schopler, Galinsky, and Abell 1997). It is unclear how common online support groups may be today. They can occur "simultaneously," meaning that members can gather via their home computers at the same time to communicate with each other in real time, or "asynchronously," meaning members can read and send messages at their convenience (Schopler, Abell, and Galinsky 1998). The benefits of online groups include increased accessibility, convenience, anonymity, and improvements to the group process. The anticipated problems or disadvantages are inability to discern interpersonal cues, technological issues, group process issues, and professional and organizational concerns (Schopler, Galinsky, and Abell 1997). Of course, online support groups are limited to the written word and the ability of participants to communicate in writing. Advances in computer technology, such as affordable videoconferencing, may alter how online support groups are used in the future.

Similarly, an organization can efficiently connect geographically dispersed and isolated workers, such as those in very rural areas or practitioners scattered across the country. One technique is computer conferencing, which can be used to create "virtual meetings." For example, it has been used to connect counselors from a university-based program who are scattered in more than seventy manufacturing plants thoughout the country (Root 1996).

INFORMATION RETRIEVAL

With the increased availability of personal computers, CD-ROMs, and modems, practitioners can access bibliographic databases to stay up to date on the latest research findings. Even before the growth of the World Wide Web, Leonard Gibbs (1990) encouraged practitioners to use such databases. However, while information may be power, it can also be a source of confusion. How can a practitioner make sense of research findings that are ambiguously stated, proceed when there are few studies that address the specific problem, or use the findings when the abstracts alone do not provide enough information to show if the study is appropriate? Gibbs suggests that meta-analyses are one possible solution and explains how they are constructed, with the hope that practitioners can make practical use of them. We know little to date about how many social workers have access to bibliographic databases and how frequently they actually use them. We do know that the research literature, whether printed in journals or available through cyberspace, is not easily digested or used, and that conducting meta-analyses is hardly within the time, skill, or analytic capacities of most practitioners (see chapter 8 on research utilization).

Nevertheless, instant access to relevant information is likely to affect the practice of many professions in the future. *Newsweek* magazine (Cowley and Underwood 1999) recently reported about the use of InfoRetriever, a small palmtop PC software program that allows physicians to enter a few characteristics of a patient and the health problem and instantly retrieve specific evidence-based recommendations regarding treatment. It has the potential to revolutionize medical decision making and practice. Unlike the sprawling Medline database of millions of articles, InfoRetriever focuses specifically on patient care and capitalizes on available digests of medical findings, literally putting treatment recommendations in the hands of practicing physicians.

Gary Holden at New York University has developed the World Wide Web Resources for Social Workers page (www.nyu.edu/socialwork/wwwrsw), which provides links to more than 50,000 sites containing relevant journal articles and government reports (Holden, Rosenberg, and Meenaghan in press).

LIBRARIES AND PUBLISHING ON THE INTERNET

Organizations that have always been in the information business, such as libraries, have been transformed. Card catalogues, the unsifted ore for the training of every scholar over 40 years old, have disappeared. As of 1996, 23 percent of the libraries serving areas of 100,000 population or more provide Internet access to their holdings, and 70 percent allow consumers to conduct database searches with staff assistance (Spencer 1999). The Internet is also beginning to affect the nature of scientific publication and distribution. The scientific article, published only after a lengthy process of peer review, bound in journals, and found in research libraries has been the cornerstone of knowledge development and dissemination in most disciplines and professions. In 1999, Harold Varmus, the director of the National Institute of Health, touched off a vigorous debate in the scientific community when he proposed that research findings be distributed on the Internet prior to or even without later publication in journals (see, for example, Pear 1999; Kiernan 1999a). His proposal for what he called E-biomed troubled many journal editors and scientific societies, both because it might undercut their market of subscribers and because it might remove one of the barriers to the worldwide distribution of bogus, shoddy, or flawed research—competent peer review. Despite the opposition, NIH proceeded with its plan (Kiernan 1999b).

SOCIAL WORK EDUCATION

Technology, in the form of audiotapes, videotapes, computers, fax machines, e-mail, television, and the Internet, is increasingly being used in social work education. A recent issue of the journal *Research on Social Work Practice* (July 2000) focused on technology and social work education, including articles on multimedia, the Internet, Web pages, chat rooms, and interactive television. Although it is unknown how widespread these technologies are in education, their use is clearly increasing (Conboy et al. 2000; Kaye 1991; Patterson and Yaffe 1993).

For example, computer technology was central to one teacher's approach to case-based learning. To address the need for inexperienced social work students to develop and practice problem-solving skills, Visser (1997) developed a case-based database that he used as part of the field curriculum. He had two goals: to create a more systematic way of teaching about cases and to create a much-needed database that could be used by future students. Each actual case entered into the database had to contain the same elements: overview, assessment, analysis, strategy, actions, and evaluation. One of the advantages to this kind of database is that it contains real-life situations, as well as the steps students went through to work with them. The major difference between this and the written teaching cases that have traditionally been used in social work education is that the database allows students and practitioners to search rapidly for cases or elements similar to what they are working on (Visser 1997). The potential advantages are that the database could link students, practitioners in the field, and university teachers in a communications network. In such a system, practitioners would have the ability to input cases that could then be reviewed by students. Universities could act as "think tanks" and interact with practitioners about particular cases. Visser (1997) anticipates students taking a more active role in their education and instructors changing the way they teach field courses in order to integrate the case-based system. Unfortunately, the full case-based approach has yet to be tested (Visser 1997).

THE FUTURE

Although computers became much more common within social work agencies during the 1990s, their application to practice has not been nearly as rapid. Computer technology continues to be used primarily for information systems, data collection, assessment, and interviewing (Butterfield 1995), and to a much lesser extent for expert systems and decision support systems (Schoech 1995). The ways computers and the use of the World Wide Web will influence social work practice during the next ten years can only be speculated on.

How will the Internet affect the ways agencies provide services and clients find community resources? Will e-mail between practitioners and clients become a method of establishing therapeutic relationships and providing help? Will clients turn increasingly to nonprofessional chat rooms and Web-based self-help groups? Will assessment, treatment planning, and outcome evaluations be conducted online with clients? Will managed care companies position themselves between practitioners and clients as the supervisors of social workers? Will clients seek out helpers with prominent Web sites and make geographical location increasingly irrelevant? The next decade will be interesting indeed.

Lessons and Lingering Issues

As agencies and practitioners struggled to adapt to these technological developments, they learned and relearned valuable lessons about how to successfully introduce innovation into social work practice. The lessons are commonplace, but they are often repeated in the social work literature. Foremost among them is the need to involve staff in the implementation process.

STAFF INVOLVEMENT IN IMPLEMENTATION

Implementing computer technology in agencies was problematic. Apathy, hostility, and skepticism were common initial reactions from staff (Benbenishty 1989; Boyd et al. 1981; Rimer 1986). Only by soliciting staff reactions were researchers able to identify effective means to counter them (see Monnickendam 1999 for a review). The implementation process itself was used to counter resistance, with varying degrees of success. Computerization called for entirely new methods of gathering, recording, storing, and retrieving information. Designing these systems required personnel and skills not available in most agencies. Ignorance and fear of computers existed from the top levels of management down to front-line workers. The latter may have feared for their jobs; the former sometimes sabotaged information systems to hide deficiencies (Schoech and Arangio 1979). In the early literature there are reports of lessons learned by the advocates of computer technology. For example, an article by Elliott Rubin, "The Implementation of an Effective Computer System" (1976), describes how one agency overcame barriers and successfully implemented a statistical information system. Three factors were key to success: no staff could be fired as a result of implementation; the computer system had to be acceptable to the staff and board; and costs had to be within reasonable limits. Staff were involved extensively in the development of this system and were enthusiastic about how it saved them time.

Schoech and Schkade (1980) reached a similar conclusion in their implementation of a decision support system. They recognized that caseworkers viewed computer technology as a burden, but if they were involved in its development, the system would be more likely to be useful rather than burdensome (Schoech and Schkade 1980). In another study, of the implementation of a computer model for use in child placement decisions, Schwab et al. (1986) worked with fifty social workers to develop the system. Supervisors at the agency insisted that the system be easy to learn and use, supportive of child placement decision making, readily available, and faster than current procedures (Schwab et al. 1986). Staff involvement was a key factor in another study in which employees felt they had been left out of the development phase but eventually accepted a decision support system after they were included in the creation of assessment forms and guidelines (Rimer 1986). A "bottom-up" (versus the normal "top-down") approach was credited with successful implementation of a decision support system in which researchers opted to work first on a small scale with the most enthusiastic staff, rather than proceeding with a full-blown implementation (Boyd et al. 1981).

Just as it was important to introduce a computer system to only the most enthusiastic workers first, a step-by-step approach appeared practical for introducing any computer application, allowing staff to become accustomed to it in phases. This was described by Rimer (1986) in the implementation of a decision support system to assure equitable distribution of resources for an in-home support services program. Implementation was conducted over a period of three years in planned phases. Resource allocation to clients was determined to have greatly improved afterward (Rimer 1986). Phasing in and careful planning are the best defenses against staff resistance to a new and threatening computer system (Bostwick 1983). Successful implementation must be incremental, modular (meaning parts of the system can be implemented one at a time), and developmental (allowing for staff input to adapt how the system operates) (Benbenishty 1989). This is basic advice for introducing any innovation in social agencies.

While computer use spread from large governmental to small nonprofit agencies and from the workplace to the home, old concerns about how it could negatively affect social work practice and clients continued to linger. We will next consider concerns about confidentiality and dehumanization.

CONFIDENTIALITY

As soon as computers were introduced into social welfare, social workers became concerned about whether and how client confidentiality might be compromised. Even those who embraced the use of computer technology in agency practice worried about protecting confidentiality (Abels 1972; Boyd et al. 1978; Noble 1971; Vondracek 1974), noting that whether a case file is computerized or hand-written and stored in a drawer, the key to confidentiality remains the ethics and professionalism of the personnel hired as direct service providers.

In an article in *Social Work* in 1971 entitled "Protecting the Public's Privacy in Computerized Health and Welfare Information Systems," John Noble takes the most conservative stand. He anticipates that privacy and confidentiality will suffer from the political motives of executives, the ingenuity of managers, and the carelessness of technicians as they get their hands on information systems. He cites an episode in which front-line social workers in San Francisco battled against computers by refusing to input client identifiers, such as name, social security number, and diagnosis, until they were assured that the Department of Social Welfare would not use this information for eligibility purposes across departments. If confidentiality is imperiled, he suggests, it will destroy the professional-client relationship that is the core of effective service. In contrast to advocates who wanted to use information systems to promote services integration, Noble is concerned about the ethics of sharing identifying information about clients with other agencies. He recommends that social workers lobby to create legislation that regulates computerization, and he feared that in thirteen years—the year of George Orwell's "Big Brother" (1984)—there would be major campaigns among social work and health care professionals against privacy infringements due to computers.

There are sound reasons why both advocates and opponents of computers have been concerned. Confidentiality is thought to be the foundation of the therapeutic relationship (Gelman et al. 1999) and historically has remained a basic tenet of social work practice (Rock and Congress 1999). Clients expect that the information they share will be kept private and not be used for purposes unknown to them (Gelman et al. 1999). There are two types of confidentiality: absolute and relative. Absolute confidentiality means that *no* information will be revealed under *any* circumstances without the informed consent of the client. For practical purposes, however, there is only relative confidentiality because of court rulings (as in *Tarasoff v. Regents of the University of California* [1974], 18 Cal. Rptr. 129, where a client revealed to a therapist an intention to harm a third party) and because of mandated reporting in cases of apparent domestic violence (Gelman et al. 1999; Rock and Congress 1999). The role of the social worker determines the circumstances under which certain information may or may not be kept confidential. It is his or her responsibility to clearly explain the limits of confidentiality to clients at the outset of the relationship (Gelman et al. 1999).

The concerns of direct practitioners about confidentiality did not wane throughout the 1980s, despite advances in the technology that allow for password protection and restricted access to information. New concerns emerged, such as who will get reports that are generated, who has access to computers, and who will see client identifying information (Sullivan 1980). The networking of computers raised questions about maintenance of private records and potential misuse of client information, as in a controversial computerized client tracking system that allowed client records to follow chronically mentally ill clients from one agency to another (*Practice Digest* 1983). Confidentiality remained among the top three concerns of social workers into the mid-1980s (the other two were dehumanization and clients receiving less service) (Pardeck 1987; Rimer 1986).

With the advent of password protection and restricted access and increased familiarity with the workings of clinical information and computerized assessment systems, social workers' fears for confidentiality were somewhat assuaged (Benbenishty and Ben-Zaken 1988; Nurius and Hudson 1988). But Finn (1988) found that only 40 percent of agencies with computers actually used password protection, and an additional 21 percent did not use password protection or eliminate client identifiers. Thus, as they get familiar with computers, workers may become desensitized to threats to security, placing confidential information in jeopardy, particularly as computer systems become networked and tied into Internet systems. The technological capability that has mushroomed in recent years has raised new concerns regarding confidentiality (Gelman et al. 1999), reinforced by legislation such as the Health Insurance Portability and Accountability Act (1996), which allows for "universal patient identifiers" so that all of a patient's records can be linked nationwide, without the patient's consent (Gelman et al. 1999). Now, as Gelman et al. (1999) note, many Fortune 500 companies check the health records of prospective employees. Even law enforcement agencies have increased access to health records (Gelman et al. 1999).

Despite questions about confidentiality since the beginning of the computerized age, the NASW *Code of Ethics* did not directly address confidentiality issues for electronic media until 1996 (Rock and Congress 1999). Social workers within managed care have a unique set of concerns. Managed care companies expect a "boilerplate clause" in insurance policies that allows them full access to patient records at any time during treatment. It is the responsibility of the helping professional to inform clients about the required access to records by the HMO and to have clients sign an informed consent to permit this, or treatment may not be provided. Many patients are unaware and uninformed of their privacy rights in the managed care environment, and they do not understand the consequences of breaches of confidentiality (Rock and Congress 1999).

The process of sharing information with managed care companies also presents problems. The common practice of faxing or e-mailing records for immediate review rather than sending them by more traditional methods leaves patient data open to the purview of whoever picks them up at the machine. Voicemail also creates concerns, particularly when client names are used. Rock and Congress (1999) propose a framework for preserving confidentiality in managed care by assigning information to three levels based on potential damage that its revelation could cause to clients (Rock and Congress 1999).

Social workers who communicate via e-mail may not be aware that all messages can be viewed by outside parties, especially when a commercial Internet service provider is used. Use of client names and identifiers must be avoided unless safety precautions are taken. Even an "intranet," a network within a large organization, can be "hacked" into unless appropriate firewalls or other security protection methods are in place. Hardware and software solutions can protect electronic information, but they are probably not routinely used in agencies. A higher level of security can be achieved by using digitally signed e-mail or by actually encrypting e-mail and attachments so that if a third party does access the documents, they will be essentially unreadable (Gelman et al. 1999).

More stringent controls for confidentiality are necessary, particularly as electronic communication quickly outpaces current laws. Until such controls are in place, Gelman et al. (1999) make several practical recommendations to social workers about record keeping. Essentially, practitioners must be careful about what they document in client records, understand who the audience of the records will be, and provide only information that is pertinent. Informed consent from clients should always be obtained. Safeguards that protect electronic data should be put in place, whether passwords or encryption.

ETHICS

Computers are now capable of functions far beyond simple data storage and retrieval, even if those advanced or "second wave" applications, such as expert systems, therapeutic games, and decision-making programs, are not yet widely used in practice. But with these technical advances come new ethical concerns. In Nurius et al.'s (1991) survey of agencies, security risks and ethical considerations were ranked second and third on a list of concerns about computer use. Among the issues raised are questions about whether expert or decision support systems may be taken as the final word in decision making, even when a client prefers a different course of action than that recommended by the software; whether smaller agencies and their clients are at a disadvantage when they cannot access online resources or when the costs of computerization come at the expense of other needed client services; and whether individualized treatment is threatened if computer systems limit what information is deemed relevant and gathered, and what treatment options are presented (Cwikel and Cnaan 1991). Murphy and Pardeck (1992) worry that the client's view of the problem might be ignored if it does not fit neatly into a standardized information system, since expert systems rely on what the "average" client looks like. Clients with unique characteristics may not be readily identified or served (Cwikel and Cnaan 1991). Assessment and treatment plans may not consider the more ambiguous components of a client's situation because they cannot be easily fitted to the information system (Murphy and Pardeck 1992).

Some authors raise other ethical concerns, such as that computer systems may promote professional elitism and separate the social worker from both the client and the client's community. They fear that social workers may ignore or avoid personal interaction when it is easier to access information on computer networks (Cwikel and Cnaan 1991) and may rely more on computerized expert systems and quantitative data than on all other types of information (Murphy and Pardeck 1992), imposing an insidious form of social control that rests on the scientific process and the experts without considering the practitioners' or clients' experiences. Finally, with so many aspects of the therapeutic process—such as assessment, therapeutic games, and some phases of treatment—now doable via computer, there are fears about whether practitioners will spend less time with clients in face-to-face interaction (Cwikel and Cnaan 1991). It is too early to tell whether this second wave of technology will actually improve or degrade the quality of services. Few studies have examined its impact.

DEHUMANIZATION

As some social workers were attempting to understand how computer technology could be usefully adapted to advance social work practice, others launched a broad attack on it. John Murphy and John Pardeck, in "Computerization and the Dehumanization of Social Services" (1992), made the most multifaceted criticism of computerization, seeing few or no benefits and much potential harm to practice. Similar issues are raised in other articles (see Pardeck and Schulte 1990; Pardeck 1997; Pardeck 1998). In making the case for "epistemological pluralism," they argue that computers are not value-free tools but are predicated on certain biased assumptions about reality that will misinform practitioners, leaving clinical interpretation and common-sense knowledge by the wayside. They worry that computers will distort practitioner-client communication and eliminate intimacy in helping relationships. They fear that practitioners will be lulled into accepting a reductionist outlook and experts will gain too much social control over clients, eliminating the humanistic tendencies that sustain service delivery. There are echoes here of Teicher's "casework is art" article from 1967, only with contemporary, postmodernist elaboration.

While most commentators acknowledge that computers could degrade the client-practitioner interaction, some are willing to note that well-designed computer technology can enhance this relationship, as in the case of a well-constructed assessment tool that allows the practitioner to become aware of a wide range of client concerns in a faster and more efficient manner (Finnegan et al. 1991; Mattaini 1993). Although computers have been criticized for their inability to make proper decisions on behalf of clients, traditional methods of making clinical decisions are certainly not error free, as Finnegan et al. (1991) point out. While this is not an excuse for inadequate decision-making models, it suggests that technology should not be held to a standard higher than that which human decisions makers must meet.

Computer use is no panacea, and Dan Finnegan, André Ivanoff, and Nancy Smythe (1991) call for guidelines to control the quality of computer technology. For responsible use, they suggest such steps as establishing a policy statement identifying how computers will be used in social work practice, having software approved by a professional social work body, and establishing a centralized clearinghouse for software related to social work. Until these elements are in place, however, they recommend that practitioners be cautious and carefully analyze software by looking at the documentation, user friendliness, aspects concerning client privacy, and means by which the program was validated (Finnegan et al. 1991).

Conclusions

Social work has not been as easily affected by technology as medicine or engineering. Its methods have remained humane but common—a compassionate expression of concern, an offer of personal help, a supportive human relationship, and proffered advice. Historically, social workers have tended to see advances in technology, which promoted industrialization and produced all manner of social dislocations and human problems, as a noxious cause for which professional social work was a partial palliative.

Thus, computer technology appeared in social work as foreign, strange, misunderstood, and anxiety-provoking. Social work generally has no animosity toward imported ideas. In fact, it largely consists of them—scientific charity, the disease model, psychoanalytic theory, various psychiatric theories of personal dysfunction, cognitive-behavioral intervention, social ecology, social constructionism, and many others. A case could be made that one of social work's strengths is its ability to identify potent ideas and adapt them to its professional purposes. But most of social work's imported products have been *ideas*, that is, cognitive formulations, nurtured in adjacent fields. Similarly, the profession's attempts to make practice more scientific have borrowed from other fields. Mary Richmond's rigorous case studies are modeled on medicine and law; research utilization was derived from attempts to improve agriculture and public health in developing countries; the idea of the scientist-practitioner was modeled on developments in clinical psychology; design and development borrowed from business and industrial engineering.

Computer technology, in contrast, was different in two respects: it came from distant fields (e.g., electrical engineering and artificial intelligence) and it arrived in the form of machines, not ideas. Computer technology promised to count and sort faster, but social work did not view itself as in the business of counting or sorting, so the relevance of the newfangled machines was hardly apparent. Furthermore, the idea of trying to quantify or make formulaic what social workers did ran against the professional grain.

The profession's researchers and scholars, on the other hand, *were* in the business of counting and sorting; those are the building blocks of science. In their view, such capabilities were needed not only to advance knowledge in social work but also to meet the external challenges of the era of accountability, which required that the profession document what it did and what were the effects. For this relatively small group of academicians, computer technology promised to make both scientific tasks easier and to assist in improving the practice of social work. Thus, as with other innovations, a few professors became the first recruits in the technological revolution.

It has been largely a revolution imposed from without, not fought and won within social work. The stimulus for adopting computers came first from government and funding bodies, which in seeking better means of oversight mandated the use of information systems. Later, as the PC revolution swept commerce and communication, computer technology made its way into social agencies and universities in one form or another, first as management information systems, then as word processing programs, and then as e-mail, promoted not so much by academicians as by the overwhelming experience of efficiency, cost savings, and the need for accountability.

Furthermore, the most common uses of computer technology in social work are not at their core part of the scientific agenda of researchers. Rather, they are driven by the organizational reality of needing to do more with fewer resources. An example is close at hand. Thirty years ago, faculty in most schools of social work shared the time of a secretary hired to answer phones and take messages, type and revise manuscripts, handle written correspondence, copy course materials, and so forth. Today, few faculty have any secretarial services. Instead they have computerized voicemail systems, personal computers with word-processing software and laser printers, high-speed copiers, and Internet connections for e-mail. As undeniably useful as this revolution has been in the workplace, it was not brought about by scholars seeking truth but by managers seeking savings. Management information systems, word processing, and e-mail do not make writing or teaching more scientific, however much they facilitate research.

Computers were at first a means of achieving organizational efficiencies, not a means of improving direct practice. What administrators desire and expect from computers has often been quite different than what would be useful to practitioners. This has at times produced conflict and resistance to the integration of computer technology. Often what practitioners might need to know is simply not included in most information databases. Jeanette Semke and Paula Nurius (1991) refer to this as a "problem of poor fit" between the information that is collected and how it is used. They suggest creating a "structure of information" that merges the needs of automation, practice evaluation, and service effectiveness. This has been the promise of computer technology since its first appearance in the social work literature.

The advocates of computer technology have seen a substantial gap between what is technically possible and what is actually widely used in social work agencies. There was indeed a poor fit. The technology itself was foreign to social work, and considerable resources of expertise and cash were required to design, develop, and constantly update information systems for agencies. But even when administrators were ready to adopt the new technology, social workers were not necessarily inclined to use it for direct practice purposes. Extensive and expensive efforts to develop expert and decision support systems rarely moved beyond the creation and initial testing phases into common use in social work. However, attempts to have practitioners use computerized client assessment instruments and client databases were more encouraging. Word processing and e-mail have rapidly become staples of practice activity, but the use of computers for treatment purposes has not yet moved beyond the experimental phase. And it is too early to know what effect widespread Internet access will have on practitioners or clients.

Both Morton Teicher, who feared technology replacing the art of casework, and Ivan Vasey, who anticipated using computers to improve agencies' efficiency and services, would find many examples in the ensuing thirty years to support their case. Teicher would no doubt be repelled by the development and use of client databases, expert systems, and computer-assisted interviewing. On the other hand, he might take solace in knowing that use of this technology among practitioners is limited, that social work still rests on a good dose of intuition, instinct, and tempered faith. Vasey would be pleasantly surprised to see the extensiveness of management information systems in agencies both large and small and the user-friendliness of the new technologies—no more entering data on key-punched cards. He might be disappointed, however, by the fact that these data-processing technologies have not stimulated practitioners to initiate research studies in their agencies.

In four decades, the profession has both overestimated and underestimated what computers might do for practitioners. It expected greater willingness on the part of managers and practitioners to adopt unproven technology, greater ability to develop sound decision support and expert systems, and easier introduction of innovations into agency practices. On the other hand, few anticipated the speed and ease with which word processing, e-mail, and the World Wide Web would transform communication. At the moment, the complete story of computer-assisted practice cannot be told. There is reason to believe it may not have even begun yet.

CHAPTER SEVEN

Research-Based Practice

Historically, as we have seen, the influence of science on direct social work practice has taken two forms. One is the use of the scientific method to shape practice activities, for example, gathering evidence and forming hypotheses about a client's problem. The other is the provision of scientific knowledge about human beings, their problems, and ways of resolving them, expressed through undertaking studies that might inform practice decisions, borrowing the results of scientific work from other fields, and creating empirically validated practice approaches. This kind of influence has culminated in what has come to be known as "empirically supported," "evidence-based," or "research-based" practice. We shall use the latter term. In our definition, research-based practice (RBP) refers broadly to practice that is informed by the results of available empirical research, that is, research conducted to build general knowledge or practice methods, in contrast to research specific to the case at hand. Although RBP is an essential part of scientific practice (see chapter 4), it can be used without that form's other components.

RBP is a complex construct with several distinct dimensions. One has to do with the extent to which practice is guided by the results of research. For example, some practitioners may use them only sporadically; others may use them consistently. Another dimension relates to the type and amount of research on which practice is based. Contrasting examples are an intervention whose outcomes have been tested in several rigorous controlled studies and one that has been tested in only one study with a number of design flaws. A related contrast sometimes made is between research that tests the *efficacy* of an intervention (highly controlled randomized experiments that give narrowly applicable results) versus research that tests the *effectiveness* of an intervention (studies with less rigorous designs that test the utility of an intervention in ordinary practice) (Chambless and Hollon 1998). (Since it is often difficult to make this distinction, we have, unless otherwise noted, used these terms interchangeably.)

Still another dimension relates to the degree of fidelity with which the practitioner applies the knowledge actually produced, which in turn is connected to the robustness of the knowledge. For example, some tested interventions may retain their effectiveness despite considerable variation in how they are used; others may need to be used precisely as described in order to be effective. A final consideration concerns the practitioner's knowledge of the research on which the intervention is based. For example, a practitioner who uses an intervention produced by a research process can be said to be doing RBP, even if he or she knows nothing of the underlying research. However, such knowledge may enable him or her to use the intervention in a more judicious or discriminating manner. When these dimensions converge at their "high ends," clear and strong examples of RBP result—practitioners make extensive and faithful use of findings based on rigorous research, with awareness of what went into the research that produced them. However, most RBP falls considerably short of this ideal.

As previous chapters have suggested, progress in attaining even low levels of RBP has been slow. A critical form of research—experiments to test the efficacy of interventions—did not appear until the 1960s, and these early efforts more often than not failed. Although the negative results served to raise doubts about the practices tested and spurred the development of new approaches, they did little to establish what might be effective. Not until the 1970s, with the acceleration of the scientific practice movement, did social work experiments begin to produce largely positive results, that is, to demonstrate that certain interventions were in fact effective means of helping clients resolve their problems. In chapters 4 and 5, we discussed how this movement produced the practitioner-researchers and research methods needed to establish a base of demonstrably effective interventions. In this chapter we will begin with these results and other research that has begun to form an empirical basis for social work practice, then take up a number of issues related to the development and use of RBP.

The Emerging Empirical Base for Social Work Practice

During the past two decades a number of reviews and meta-analyses have identified a sizable body of demonstrably effective practice methods used by social workers (deSchmidt and Gorey 1997; Gorey 1996; MacDonald, Sheldon, and Gillespie 1992; Reid and Fortune, 2000: Reid and Hanrahan 1982; Rubin 1985; Sheldon 1986; Gorey and Thyer 1998; Videka-Sherman 1988). (A meta-analysis is a synthesis that combines the data of other studies, in this case usually controlled experiments, testing the effectiveness of different types of intervention. See for example Rosenthal 1984; Fischer 1992). The most recent review (Reid and Fortune 2000) provides current detail about the nature of these methods, identifying all empirically tested programs in social work reported in the literature during the 1990s. In order to be included, a program had to have been tested through a design that was either experimental (randomized) or quasi-experimental (using a non-randomly formed comparison or control group). Programs tested by controlled single-system designs were also included, if they involved six or more clients. Further, the program evaluation had to be reported in a publication in which at least one author was a social worker or affiliated with a social work school or other organization. A total of 129 programs met these criteria; the majority used randomized designs. The programs occurred in all major fields in which social workers practice. Most frequently addressed, in this order, were problems of mental health, child/youth behavior, substance abuse, aging, health, domestic violence, and child abuse/placement. For the overwhelming majority of the programs (88 percent), evaluations revealed positive findings on at least one major variable.

The dominant modality was some form of group program (59 percent), typically small groups. Programs centered on the individual constituted 20 percent of the sample, and family programs about 9 percent. The majority were short-term. For 63 percent, the duration of intervention was 12 weeks or less. Only 12 percent lasted longer than one year. The programs emphasized use of group modalities, a high degree of structure, planned short-term designs, cognitive-behavioral methods, and formal instruction, such as education, didactic presentations, and training modules. About 10 percent used case management methods, especially with the seriously mentally ill and families with children at risk of placement.

These methods are part of a much larger set of interventions of proven efficacy available to the helping professions. For example, Reid (1997a) reviewed 42 meta-analyses (in 31 problem areas) that examined the results of several thousand experimental tests of interventions in the helping professions that are (or could be) used by social workers. The vast majority of studies in these meta-analyses reported positive effects. Although behavioral and cognitive-behavioral methods predominated, positive effects were found for a wide variety of approaches. Certainly an ample body of research-based methods exists for many areas of practice, especially for well-defined problems such as anxiety, depression, and child behavior difficulties.

The emergence of this body of tested practice methods has led to development of criteria for defining empirically supported methods. The most extensive work has been carried out by the American Psychological Association. Its efforts, which have been extended in recent work by Chambless and Hollon (1998), use the quality of research evidence as a basis for nominating certain interventions as empirically supported or their language as "efficacious" (following the efficacy/effectiveness distinction noted earlier). As discussed in chapter 5, criteria include the use of rigorous randomized designs and replication of findings in studies by at least two independent research teams.

Others have applied these criteria to different fields of practice to determine which types of treatment are efficacious. Some examples are: cognitive and interpersonal therapies for depression; exposure for agoraphobia; exposure and response prevention for obsessive-compulsive disorder; education and behavioral exposure for child anxiety; cognitive problem-solving skills training as well as parent management training for oppositional and aggressive children; behavioral marital and emotion-focused therapy for marital conflict.

These results, as well as the meta-analyses cited earlier, show that behavioral and cognitive-behavioral approaches appear to dominate. But other forms of intervention have also established strong empirical credentials. For example, among the efficacious interventions mentioned above were interpersonal therapy for depression and emotion-focused marital therapy, both of which use expressive and insight-oriented methods. Evidence has accumulated of the effectiveness of certain models of short-term psychodynamic intervention for depression and other emotional problems. One reason for the current preeminence of behavioral and cognitive-behavioral methods in RBP is that much more of an effort has been made to validate those types of methods.

The growth of research-based interventions has been matched by increases in research-based knowledge related to assessment and human functioning. Such knowledge provides empirical grounding for assessment as well as for theories underlying intervention methods. For example, in their work with parents and children, social workers make use of research-based knowledge on norms and variations in human development and behavior, such as when infants and children, on the average, reach developmental milestones (Bloom 1985), as well as knowledge of the effects on children of different kinds of trauma, such as divorce (Zaslaw 1988, 1989). Although research has produced few fully developed explanations of problems of psychosocial functioning, it has identified risk and protective factors. Problems for which study of such factors has produced knowledge of direct value to practitioners include juvenile delinquency and drug use (Smith et al. 1995), schizophrenic relapse (Hogarty 1993), and suicide in special populations (Brent 1995; Ivanoff and Riedel 1995).

Other examples can be found in recent research in psychodynamic and humanistic schools of practice. This research is more likely to consist of study of underlying constructs and treatment processes than conventional effectiveness testing. Some examples: Goldstein (1998) used studies of female development to correct traditional object relations theory, with clear implications for work with lesbian clients; Wolmark and Sweezy (1998) used recent research on infant behavior to reformulate some of the foundational assumptions of Kohut's self-psychology. These examples, as well as the emergence of research-validated interventions noted earlier, suggest that psychodynamic and humanistic approaches are themselves slowly moving in a more empirical direction.

Results of such research take the form of probabilistic generalizations. Given the risk factors for a client with certain characteristics, how likely is it the problem will occur? Although the probabilities can seldom be quantified, they can be used as a basis for preventive or other action.

Issues

A substantial amount of tested intervention and assessment knowledge has already accumulated, and this base continues to grow. However, it is unclear to what extent this can usefully inform social work practice. We turn now to some of the central issues related to research-based knowledge as a foundation for practice, focusing on the use of interventions whose effectiveness has presumably been demonstrated through research.

THE SOUNDNESS OF THE BASE

As noted, many social work intervention programs can claim some measure of effectiveness based on research evaluations. But how good are these empirical credentials? The hallmark establishing the effectiveness of any intervention—independent replication—has seldom been done. Further, most experiments supporting the efficacy of interventions in social work and related fields have been carried out by practitioner-researchers with a stake in the interventions tested—they either originated them or had some prior convictions about them. Such involvement can result in a serious and pervasive form of bias—what Rubin has referred to as "expectation of improvement" (1999:637). "Experimental demand" and "investigator allegiance" are among the more widely known terms used to describe this phenomenon. We shall use the latter in the following discussion.

Investigator allegiance is a form of bias in which the outcome of an intervention experiment is shaped in the direction of the researcher's expectations, hopes, and other predictions. There is a good deal of evidence to suggest that such effects occur (Smith et al. 1980; Robinson et al. 1990; Gorey 1996). This is a problem because most intervention experiments are conducted by adherents who have a stake in the outcome. For example, in the Reid and Fortune (2000) review, 90 percent of the evaluations of programs found to have positive results had been conducted by either developers of the interventions or researchers who appeared to support them. Moreover, investigator allegiance is more likely in research-based approaches since adherents are more likely to have research skills and interests than practitioners of a more humanistic persuasion, and thus are more likely to test their own or similar interventions.

However, there are some difficulties in the notion of investigator allegiance. As Weiss and Weisz (1995) argue, expectations that a given method is effective may be based on prior evidence of its efficacy and not just belief. Thus, ruling out allegiance may also rule out evidence. For example, in experiments comparing an intervention that has been repeatedly found to be *ineffective* with some rival intervention that has shown promise in experimental tests, most researchers would expect the rival intervention to prove superior since that is what prior evidence has suggested. Thus they would be likely to make their expectations known in the research report and would be seen as "identified" with the intervention they tested.

Difficulties in defining investigator allegiance also must be taken into account in replications of tests by independent research teams, a central criterion for efficacy in the work of the American Psychological Association, discussed earlier. For example, most interventions with enough research support to be considered efficacious are behavioral or cognitive-behavioral, and independent teams replicating original studies may well have similar orientations. Thus even independent replications might be considered vulnerable to investigator allegiance.

Be that as it may, this form of bias remains a complicating issue in evaluating the empirical credentials of an intervention. Some solutions, however, are possible. For example, when alternative interventions are tested against each other, advocates of both methods can be included on the research team.

USEFULNESS OF RESEARCH IN GUIDING PRACTICE

A second issue is the value of empirically based knowledge in guiding practice. Not everyone agrees that research is a better basis for selecting interventions than other sources of knowledge, such as practice wisdom. Witkin (1991a) has argued that lack of objective criteria for determining empirical validation and limits on generalizing findings nullify the presumed superiority of research-based methods. Chandler (1994) has also questioned the utility of RBP, given the typical lack of correspondence between supporting studies and the clinical situations faced by practitioners.

The lack of consensus about when the effectiveness of an intervention has been established is, of course, a limitation on the use of RBP, but not a valid reason for not using it at all. The same limitation applies to any sort of empirically supported intervention in any discipline, or, for that matter, to any sort of scientific finding. There are no universally accepted criteria for establishing truth (chapter 1). Certain guidelines, like those of the American Psychological Association, can be suggested, but not everyone will agree with them. The case for the superiority of RBP over other forms does not rest on "objective criteria" but rather on the assumption that the majority of social work researchers and practitioners can agree that in fact some forms of practice have better empirical support than others. For example, the well-established empirical grounding for exposure therapy as the most effective method for treating many phobias has probably convinced most leaders of the practice community—those responsible for training and practice programs as well as those engaged in direct practice—that that form of therapy is preferable to therapies based solely on interactions between the client and a clinician (Luborsky 1995). At present the number of interventions about which such consensus can be reached is limited, but enough exist to establish the principle, and they will surely increase. Also, of course, awareness of research on the part of the practice community is needed, as will be discussed below.

Similarly, there is reason to suppose that consensus about practice knowledge or methods found to be ineffective has contributed to their gradual disuse. Among many examples are the schizophrenogenic mother, double-bind theory as an explanation of schizophrenia, psychodynamic casework as a means of helping families exit welfare, and facilitated communication with autistic children. Granted, research has not been the only factor at work, and a considerable amount of knowledge and practice persists without scientific support. Still, it is hard to find many examples of knowledge or practice that has flourished for lengthy periods in the face of repeated failures of research to support its validity.

A recent survey of practitioners' use of facilitated communication shows how the process may work and provides a dose of sobering reality. Of 177 practitioners surveyed (in North Dakota), 28 percent had previously used facilitated communication but were no longer doing so and gave it low ratings for validity and usefulness. However, despite consistently negative research evaluations, some 17 percent were still using the method and rated it highly (Meyers et al. 1998)! As the example suggests, the demise of an ineffective practice may result from a combination of its failure to produce expected results in actual use *and* negative research findings over time—an example, perhaps, of conceptual utilization of research. It also illustrates the length of time that may be involved in the decline of a practice.

There are also no objective criteria for determining the generalizability of an intervention. How do we know that a service found to be effective with middle-income clients in Los Angeles will work for lower-income clients in Tuscaloosa? Setting, geography, client and problem characteristics, practitioner qualifications, and degree of adherence to intervention guidelines are among the factors that may influence decisions about whether a tested intervention will retain its effectiveness if used elsewhere.

Ideally, generalizability should be established by replications of the intervention in different settings, with different populations. As noted, such replications are rare. However, for many empirically supported interventions there is enough corroborative evidence from evaluations of similar interventions with somewhat different client groups, practitioner types, etc., to form a logical basis for claiming that the intervention in question is likely to be generalizable beyond the limits of any given study. The review of meta-analyses referred to earlier (Reid 1997a) provides examples of numerous effective interventions similar enough to justify a synthesis of evaluations of them. Moreover, in their review of empirically supported social work programs conducted during the 1990s, Reid and Fortune (2000) found that the majority could be connected to similar tested programs in related fields. In both reviews, it was possible to locate clusters of studies supporting the effectiveness of different types of interventions and thus to build the case that these interventions had some degree of generalizability.

The evidence supporting the research-based method may make a more persuasive case for its use than whatever rationale might be supporting an alternative approach. For example, a relapse prevention program may have been found, through controlled studies, to be relatively effective with middle-class Caucasian men addicted to cocaine but have received no testing with other populations. The program could not be readily assumed to also work for lower-class African American women similarly addicted. Let us suppose, however, that there is little evidence that current programming used with the addicted African American women is effective, and that other programs that might be considered for this group have never been tested. If so, the empirically supported program may become the preferred option, since there is at least *some* hard evidence that it works with cocaine addicts. Granted, it would need to be modified in applications to African American woman through knowledge based on other research or on experience. Also, there would be the risk that it might not be any more successful than current programming. Research-based interventions or assessment theory may offer the practitioner what Fortune and Reid (1999) have called the "best available knowledge" and Klein and Bloom (1995) have referred to as the "best available information." In other words, the research credentials of an intervention can be problematic yet still enough evidence of effectiveness to warrant its use over options supported by less evidence.

IS THERE ENOUGH OF AN EMPIRICAL BASE TO MAKE A DIFFERENCE?

A third issue is whether there is enough research of sufficient rigor to provide an adequate base for practice. On this we agree with Berlin and Marsh (1993), who put it quite bluntly: "Despite the importance of empirical knowledge, it is insufficient for guiding practice. Practitioners must frequently, if not usually, use methods that lack an empirical base" (230). However, as we indicated earlier in the chapter, a sizeable body of empirically supported knowledge is currently available, enough to make RBP viable in a range of practice situations. Even if its use is limited, RBP could serve as a model of best practice (Gambrill 2000). As our empirical base grows, its role will surely expand.

An issue that has cast a large shadow over efforts to develop a research base for practice is the differential effectiveness of tested interventions. Since the 1970s a considerable amount of evidence has been amassed supporting the effectiveness of a large variety of interventions, both in social work and in related fields. In social work, at least, the cry of that era was that "nothing works." As the new century begins, it seems as if "everything works," or just about everything tested. But this apparent good news masks a major difficulty. If everything works, scientific findings provide a limited guide to practice, like a compass with arrows pointing in various directions. To be sure, practitioners might find it of value to know that intervention A had empirical support. But if comparable support could also be found for reasonable alternatives to A, such as interventions B, C, and D, then the practitioner would not get much specific guidance from the research underpinnings. What is needed is evidence pointing to one or two of these alternatives as superior to the rest, saying these interventions are the "treatments of choice."

Numerous studies and reviews have been conducted in the helping professions to determine if in fact any particular kind of intervention is superior in outcome to any other (Luborsky 1995; Weinberger 1995). But the dominant finding has been something called the "tie-score effect." Typically, quite different kinds of interventions have appeared to be effective, that is, better than control groups, but with little evidence that one was better than the others. The "tie-score effect" was dubbed the "dodo bird verdict" by Luborsky, Singer, and Luborsky (1975). (In *Alice in Wonderland*, the dodo bird declares, at the end of a chaotic foot race in which contestants could begin to run when they felt like it, "Everyone has won and all must have prizes.") As Beutler commented, "The prevailing belief, even among renowned investigators, has been that all psychotherapies are relatively equivalent (all have won)" (1991:226). This belief is still applied to human service interventions, including social work (Stalker, Levene, and Coady 1999; Greene and Jensen 1996).

We think it is time to give the dodo's verdict a closer look. First we see that it rests on rather limited and dated evidence. For one thing, it is based on mostly pre-1990s studies comparing methods of psychotherapy for adult emotional problems. For another, experiments comparing rival forms of treatment often produce tie scores, but the reason may be more lack of statistical power in the experiment than lack of differences in the effectiveness of the treatments—that is, sample sizes may be too small for differences to be detected (Kazdin and Bass 1989).

Such criticisms suggest that it is worth asking: Does research now provide evidence that can guide clinical social workers and other practitioners in their selection of particular interventions for particular types of clients and problems? This is, of course, the key rationale for using RBP—that it can help us provide better service to our clients. We believe that it can. We would like to review briefly some bases for this assertion and give some examples of intervention approaches that can be said to have better empirical credentials than others.

One kind of evidence comes from recent meta-analyses of intervention studies. Reid (1997a) examined 42 meta-analyses published since 1986. These covered thousands of studies of interventions for a wide range of problems dealt with by social workers and other human service professionals. Thirty-one, or three-fourths, of these meta-analyses reported in their conclusions the existence of differential effects between competitive interventions. Perhaps the strongest pattern is the apparent superiority of behavioral and cognitive-behavioral interventions over alternative methods in a number of problem and population areas, most notably problems of children, child abuse, juvenile delinquency, panic, and agoraphobia. Another pattern is the advantage enjoyed by multicomponent interventions (for example, a combination of rational emotive therapy and social skills training) over their rivals. The superiority of this pattern occurs across a variety of problem areas including bulimia, obesity, child internalizing disorders, DUI offenses, drug abuse, and juvenile delinquency. In other words, the meta-analyses suggest, to mix metaphors atrociously, that it may be better to have more than one arrow for your bow than to put all your eggs in one basket.

Other studies have produced additional examples of differential effects, for instance, Gorey and Thyer's (1998) meta-analysis of 45 social work studies. In their review of experimental research in social work during the 1990s, Reid and Fortune (2000) identified 31 experiments in which two or more alternative interventions were compared. Two thirds of these comparisons produced significant differential effects.

A second basis for the assertion that research can guide practice may be various problems where only one or two, or perhaps a cluster, of closely related approaches have been successfully tested and rival approaches have not been. We refer to this as the "forfeiture effect": rivals have not shown up on the research playing field. A good example is psychoeducational and family behavioral interventions with families and their schizophrenic members. These related approaches have been well tested and found to be effective in preventing relapse of schizophrenics (Hogarty 1993; Lehman et al. 1998; Penn and Mueser 1996). There are no other methods with such empirical credentials. Another example can be found in the field of corrections. Here the empirical ground has been occupied by cognitive-behavioral interventions, many of which have been found to be effective with delinquents (Andrews et al. 1990; Izzo and Ross 1990; Lipsey 1992). Some rivals prominent in work with other populations, such as psychodynamic practice, have little in the way of empirically validated methods usable with these clients.

The last kind of evidence we shall cite supporting the capacity of research to discriminate between more and less preferred methods is efficiency. Two methods may be equally effective in terms of outcome, but one may accomplish its outcome in less time or at less cost. Efficiency measures can then become ways of making research-based determinations of which interventions to use. Considerable evidence has accumulated over the years that short-term methods can accomplish as much as, in less time than, their long-term counterparts for a wide range of problems, including depression and marital conflict. Moreover, interventions can sometimes demonstrate cost savings over alternatives. A stunning example of such a study was done by Peak, Toseland, and Banks (1995). They found that support groups for caregivers of frail veterans resulted in a savings of over $7,000 per year per patient over routine service!

PACKAGING AND DELIVERING RESEARCH-BASED METHODS

Another issue of using RBP is dissemination and utilization. Critics have argued that it is impossible to keep practitioners abreast of research-based methods (Chandler 1994). In developing the necessary diffusion technology, we would do well to take note of the "evidence-based medicine" movement, which is gaining adherents in both the U.K. and the United States. Problems of practitioners' nonutilization of research plague medicine as well as social work, despite the former's vastly larger base of scientific knowledge. Evidence-based medicine is an effort to make available to physicians and related professionals research that would inform their treatments. A key component is a large-scale worldwide effort, the Cochrane

Collaboration (http://www.cochrane.org/), which makes available reviews of research evidence as to what works and what doesn't in medical practice. Its reviews in mental health areas are already of use to clinical social workers. The information is being made available in consumer-friendly form in journals, on CD-ROMs, and on the Internet. A similar initiative that will have a social work component is the Campbell Collaboration (http://campbell.gse.upenn.edu/). These efforts are harbingers of the use of the Internet as a major means of keeping practitioners abreast of developments in research relating to practice. Research reviews or other information about effective practice can be made available, and continually updated, in a form readily accessible by clinical social workers (see also chapter 6).

Another development that should accelerate the use of research-based interventions is the use of practice guidelines (Howard and Jensen 1999, in press; Proctor and Rosen in press), assessment or intervention protocols based on some combination of research findings and expert opinion. They are typically developed by panels of experts who synthesize state-of-the-art methods, giving considerable emphasis to those with research support. Practice guidelines have been employed for over half a century in medicine, with use increasing in recent years; moreover, there is evidence that they improve patient outcomes (Howard and Jensen 1999). Although their application in social work is still new, there is a good deal of interest in exploring their potential. As Williams and Lanigan have commented, "Practice guidelines are clearly the wave of the future and a very hot topic in professional circles today" (1999:338). Guidelines have recently been the focus of a special issue of a major social work research journal (see Howard and Jensen 1999) and of a conference involving leading members of the social work research community (see Proctor and Rosen in press).

Practice guidelines in some form have, of course, been used by social workers since the beginning of the profession. Indeed, Richmond's *Social Diagnosis* incorporates a very detailed set for conducting assessments. However, until the latter part of the past century, guidelines were based largely on practice theory and wisdom. Only in the last two decades has there been any appreciable use of those informed by research. For example, in their review of experimentally evaluated programs cited earlier, Reid and Fortune (2000) surveyed investigators about use of guidelines by practitioners implementing the program. In the majority of programs, written practice guidelines were used in implementing the interventions; a third of the evaluators who responded indicated that they had been implemented in other settings. (The response rate was 80 percent.) In some cases rather wide utilization was reported. Moreover, research-based guidelines are increasingly becoming a part of social work practice texts (see, for example, Lecroy 1994; Thyer and Wodarski 1998). Although such guidelines may be based on the work of a single team of practitioner-researchers, lack an extensive empirical base, and currently inform only a small fraction of practice, nevertheless they represent beginning steps on which to build.

In many respects the panel-constructed guidelines now being proposed are an ideal way to base practice on available research findings. They offer a systematic synthesis of assessment and intervention methods based on the best knowledge available, in a form that social workers can use directly in work with clients. Practitioners are not required to read and distill research reports and make judgments about the validity of the findings; instead, they use the research indirectly (see also chapter 8). By combining research-based knowledge and expert opinion, guidelines can present coherent approaches that might not be possible using only research-tested methods.

However, the development and use of guidelines must overcome a number of hurdles. As Kirk (1999) has argued, the empirical bases of many assessment and intervention approaches are thin and equivocal; practitioners may not use them if they are as lengthy, complex, and qualified as, for example, the American Psychiatric Association's sixty-three-page "Practice Guidelines for the Treatment of Patients with Schizophrenia" (APA 1997). Moreover, those setting guidelines need to grapple with the multiple meanings that "effectiveness" may have for different stakeholders, a phenomenon that has been well documented (Lambert and Hill 1994). Richey and Roffman (1999) pose an additional challenge. Guidelines in medicine and psychiatry have focused on treatment of the patient. Contemporary social work practice encompasses multiple roles beyond providing direct services to clients, such as "system linkage" (e.g., brokerage and advocacy), "system development" (e.g., creating programs), and "system maintenance" (e.g., facilitating service delivery). These roles add complexity and also interventions that often lack empirical support. Other hurdles include securing practitioner adherence. In sum, practice guidelines hold much promise and their development should be pursued. We expect, however, that progress will be slow and uneven.

Suppose we do deliver research-based interventions to practitioners, in the form of guidelines, manuals, courses, workshops, practice texts, and so on. What happens then? In an age of practice eclecticism, are social workers likely to use these methods in bits and pieces, thus nullifying their empirical credentials? As Richey and Roffman ask, "How much [can] the intervention plan be altered before it is no longer viable?" (1999:317). Treatment manuals and guidelines have the potential to keep practitioner adaptations within acceptable limits, but they by no means guarantee that this will be case. Furthermore, what "acceptable limits" might be is an important empirical question that can be examined through existing or additional research, using some of the studies referred to earlier that have produced "tie scores" between variations in interventions, at least where comparison samples are of sufficient size. For example, in their evaluation of exposure therapy for obsessive compulsive disorder, Emmelkamp et al. (1990) found that using partners to assist treatment did not change the effectiveness of the method, which had been established in other research (see DeRubeis and Crits-Cristoph [1998] for a review). Thus there is evidence that using partners in exposure would be an "acceptable" variation that would not change the effectiveness of the basic method. But more research on this point is needed, especially to determine how variations in use may alter the effectiveness of an intervention.

Use of Empirically Supported Practice and the Scientist-Practitioner

The slow but continuing rise of RBP has many implications for social work and other helping professions. We would like to comment on two. The first concerns the social work profession's view of scientifically oriented practice as a whole. We noted in chapter 2 that the influence of science on social work has expressed itself historically in two ways: the use of scientific methods as a part of practice and the use of existing research to guide practice. We have tended in recent years to bundle these two components into one inseparable package and to speak of the scientist-practitioner who uses research techniques to assess and evaluate cases *and* uses research to guide practice. These components complement each other and are important in effective practice, but they do not have to be used together. A practitioner can use research-based intervention methods without using research tools, such as assessment or evaluation instruments, and of course the converse is also true. Unbundling these components would facilitate use of RBP by those who do not want to employ research tools in their practice. A practitioner could, in effect, buy RBP more easily if it were marketed separately. Moreover, the current bundling has, we think, overemphasized the assessment and evaluation aspects of scientifically oriented practice to the neglect of empirically supported methods. Scientific tools are important in social work practice but *in themselves* can make only a limited contribution to helping clients resolve their problems. For example, careful charting of the client's progress will not be of much help if the intervention is ineffective. Scientific knowledge has the potential to create effective interventions and to demonstrate their effectiveness. It should be the senior partner. Measurement of what we are doing should be ancillary to effective practice, not a substitute for it.

The second implication concerns the promotion of RBP in social work practice and education. There have been some radical recommendations. One, by the Task Force on Social Work Research (1991), would mandate reliance on RBP in practice courses as a condition of accreditation. Another proposes that not providing a client with an empirically validated intervention when one is available is unethical practice. For example, Joe Brown, seeking help for his fears about leaving his apartment, is seen by a clinical social worker. There is abundant research evidence that some form of exposure helping Mr. Brown to leave his residence through incremental steps (almost literally!) would be effective. The social worker, however, is unfamiliar with this method and uses an approach that lacks evidence of effectiveness. Does Mr. Brown have a right to receive exposure therapy? Does the social worker have an obligation to offer it or to refer him to someone who can? Myers and Thyer (1997) would say yes to both questions. They argue that our professional code of ethics should require that practitioners use empirically validated treatment when it has been established.

We think the recommendations to use accreditation and our ethical code to further empirical practice should be considered seriously by accrediting bodies and our professional association. It is unrealistic to expect revolutionary initiatives, which may be premature in any case. But we should consider what beginnings may be made. Even if there are only a few empirically validated treatments of choice for only a few problems, we need to come to terms with the imperatives that knowledge suggests.

Much can be done on a more modest level to further the cause of empirical practice, particularly in the area of social work education. Educators can develop model curricula in RBP for practice courses at the bachelor's and master's levels, making use of the work of Calhoun and her colleagues (1998), who developed guidelines for training psychologists in RBP methods in coursework, internships, and continuing education. Social work educators can survey existing courses to make sure that available empirical content is being incorporated, just as they do to make sure that diversity content is taught. They can also work with field instructors to facilitate students' use of RBP in their internships.

If we move toward greater emphasis on RBP we will, sooner or later, have to deal with some difficult issues. For example, should the *primary* focus in practice courses be on empirically validated methods? Should expertise in and a commitment to RBP be criteria in faculty hires? Should internships be permitted in agencies not willing to allow students to use RBP? If we go this far, are we repeating our psychoanalytic past with a new ideology—RBP?

Although all such issues should give us pause, they can, we think, be resolved. RBP provides the means of putting social work practice on the kind of scientific foundation that the profession has been striving to achieve for the past century. We believe it will become an increasing force in social work practice and education.

CHAPTER EIGHT

Research Dissemination and Utilization: Spreading the Word

W hether practice is influenced by science through borrowing its technologies as methods or being informed by the findings of scholarly inquiry, the ultimate goal of using science in social work is to shape the behavior of practitioners. To be useful, science must make it out of the laboratory and into the field. For decades, it was expected that scientific knowledge as expressed in research reports would find its way into the hands of practitioners, who would use it to improve services to their clients. The need for the dissemination and use of research grew as social work matured as a profession separate from the academic social science disciplines. The demands of professionalization required that social work be linked to a knowledge base and that practitioners be trained to keep abreast of scientific developments bearing on clients' needs and effective ways of meeting them. Having divorced itself organizationally from the social sciences, social work struggled to develop an internal capacity to generate practical knowledge and to ensure that practitioners were informed about and comfortable in using it. This chapter traces the emergence and evolution of concerns about research utilization and the barriers to its success. We conclude by providing an expanded view of the nature of research dissemination and utilization in social work and suggesting directions for the future.

By the 1930s, at least some scholars recognized the need for social work to have a firmer scientific basis—common sense and good intentions were not enough to support a modern profession. In his study of social work practice (previously discussed in chapter 2), Maurice Karpf (1931) argued in favor of increasing greatly the scientific status of social work. He knew that the field needed more than practice wisdom to be viewed by the public as a profession. He promoted higher admission standards for schools of social work, improvement in the content of social work education, and increased use of the social sciences (and other disciplines) in the practice of social work. Although he clearly intended to promote a scientifically credible profession, his documentation of the underlying weaknesses in the practice and the educational programs of his day can be read as a critique of the pretensions of the developing profession.

Karpf did not advocate or expect that social workers themselves would create this scientific knowledge. Rather, it would be imported from the academic disciplines through requirements that social workers obtain rigorous study of the sciences as undergraduates and that graduate social work teachers carefully scour the social sciences for knowledge to transmit to students that could inform their professional techniques. In his view, practitioners should have knowledge of scientific methods and an appreciation for the social sciences, though not all could or would be scientists themselves. He called for a division of labor in which social scientists developed the relevant knowledge, social technologists (knowledge engineers, in current language) adapted it to practical use, and practitioners applied it in the field. These three different tasks were carefully and fruitfully linked:

The continual development and increasing complexity of social life will make necessary constant modifications of social theory and social technique. Since sound technique must depend on social science, even as social practice must provide the acid test for both, it follows that the practitioner must keep himself abreast of the developments in science for inspiration and guidance, just as the scientist must keep in touch with social life. It is obvious that it is out of the question for every practitioner to have either the necessary time or training to work out a scientifically acceptable program for the many complex situations which come to him for attention. He must turn to the technologist for help. The cooperative efforts of the social scientist seeking to discover the laws governing human life and social organization; the social technologist endeavoring to work out the practical applications of the discoveries and formulations of the scientists; and the social practitioner testing and utilizing the applications in everyday practice, would be productive of a degree of progress impossible in any other way. The three groups working together would constitute a pyramid of effort of which the discoveries of the scientists and the applications of the technologists would be the base dependably supporting the work of the practitioner at the apex.

(Karpf 1931:383–384)

The social technologists, those who design knowledge-based interventions, were a critical group in his schema. "The only hope for developing the applications of sociology and the other sciences to social work lies in the development of a group of persons who will devote themselves exclusively to this problem and will be given the freedom and the necessary facilities to pursue their interests in applications" (380). "This group may well be made up of the teachers in the schools of social work" (381).

This linking role for social work faculty would, of course, require them to be scientifically literate and better trained than those Karpf found. What he referred to as "the problem of the teacher" (373) was the fact that many social work teachers were very experienced practitioners but not necessarily ardent believers in the need for science in the profession. He worried that unless they had good scientific training, they might be skeptical or even contemptuous of science. "This is bound to communicate itself to . . . students with disastrous results so far as a scientific approach to social work is concerned" (376). Teachers might need advanced education.

The development of social work education, in fact, was generally not driven by concerns about scientific methods, competing academic theories, or debates about how to transfer knowledge. For seventy-five years, the major educational debates concerned whether social work training should be given in colleges or universities at all, whether students should be prepared for service in public or private agencies, whether they should be trained as clinicians or administrators, whether social science theory had any role in their preparation for practice, whether the training should be at the undergraduate or master's level, or whether faculty almost always drawn from pools of experienced practitioners needed doctoral education (Austin 1997). Only in the latter half of the twentieth century did leading graduate schools of social work in major research universities begin to consider seriously requiring faculty to be trained as scholars and to engage in research, and even this effort was controversial.

Although Karpf identified a series of problems that would still bedevil social work education a half century later, he could not have foreseen just how difficult the challenges would be. For one thing, much social science knowledge was not readily usable, even with the mediation of a technologist. For another, the problem remained of how practitioners, once they graduated, would continue to learn from science. Professors can only impart current knowledge to students, and its direct benefits last only if professional knowledge never changes. To the extent that knowledge evolves over time, practitioners must find a way to keep learning. Although Karpf recognized that most practitioners could not be scientists too, he did not address the full range of difficulties of bringing science into the service of practice.

How can scientific findings be shared, not only with other scientists but also with practitioners? The answer is a place where they can be stored, modified, extended, and easily accessed. Since the seventeenth century, the mainstay for spreading the scientific word has been the scientific journal. Journals preserve and document the past, disseminate information from contemporary scholars, and aid in the development of a field's knowledge base by providing a layered repository of the sediment of an era's research.

By Karpf's time, scientific and professional journals were assuming great significance in the social sciences and other fields (see Abbott 1999 for an in-depth analysis of their evolution in sociology). Journals provided a method for scientists to establish priority (i.e., ownership) over discoveries, preserve scientific contributions, and, through peer review, gain authority from having work recognized by other members of the scientific community (Zuckerman and Merton 1971). Professionals—including social workers—generally do not claim to be researchers themselves, but they project a public image of being conversant with the scientific literature, informed by the latest knowledge, and dedicated to shaping their practice on the basis of new findings. Their training, accordingly, is concerned with mastering fundamental scientific principles and methods of inquiry. With the growing importance of research and publication and the need for social work to claim a scientific grounding, it was reasonable to expect that practitioners would pay attention to research reports in journals. By the middle of the twentieth century, social work practitioners were expected to monitor the development of knowledge in the social science journals and in the emerging journals devoted specifically to social work, and to apply that knowledge in practice. The simplicity of this idea was belied by the practical difficulties, as we shall see.

The Emergence of Social Work Research

At first, "social research" was defined broadly enough to encompass not only the fact gathering, statistical reporting, and surveys conducted by social workers but also the activities of social scientists. Gradually, however, as mentioned in chapter 2, a telling distinction emerged between social research and "research in social work" or "social work research," pertaining to who did the research and, more important, to the function of the study. Social work research, early authors insisted, consisted of studies of social work practice and studies whose purpose was to advance "the goals and means of social work in all its ramifications" (MacDonald 1960:1). It is defined by the function it serves, which is producing knowledge that can be put to use. This adjacent task of research—using the results—had been relatively neglected until the 1960s. In fact, the terms "research use," "research utilization," and "research dissemination" do not appear in the index of either Sidney Zimbalist's or Norman Polansky's earlier, well-known books. While attention to research in social work was expanding, the issues of its dissemination and use were largely neglected. Both this attention and neglect can be traced in the various editions of the *Encyclopedia of Social Work* issued periodically by the National Association of Social Workers.

ENCYCLOPEDIC COVERAGE

The first edition of the *Encyclopedia of Social Work* (Hall 1965), the successor to the *Social Work Year Book*, contains only one entry on research, "Social Work Research," written by Ann Shyne. She notes that social work research "has expanded and improved substantially in quality in the past twenty years" (763) and attributes this to the profession's desire to refine its theoretical base, the increase in newly trained researchers, advances in electronic data processing, a rapprochement with the social sciences, and increased funding from the federal government. Her overview describes the private and public agencies that support research, the few research centers in schools of social work, and the major social work studies carried out. She notes that the number of schools offering doctoral education increased to fourteen and the number of social work dissertations completed quadrupled in a decade. Shyne does not mention research dissemination or utilization but does identify a potential problem:

Although the nature and quality of current social work research varies widely, such research is increasingly characterized by methodological sophistication, a delimited rather than diffuse focus, and problem formulation in terms of evolving theory and concepts rather than in terms of discrete, factual questions. These developments hold out great promise for the expansion of social work knowledge, but in the meantime *they increase the difficulty of communication between researcher and nonresearcher*, add to the duration and cost of research projects, and create some impatience in the profession and the community because of the gap between the problems needing solution and the ability of the researcher to solve them. (763, emphasis added)

The 1971 *Encyclopedia* includes three articles about "research in social work." The first, by Polansky (1971), covers the role and purpose of research and its historical origins. In describing the development of the social survey, covered by most commentators, he mentions Booth's study of the poor in London at the end of the nineteenth century, Durkheim's study of factors related to suicide rates, and the well-known study, led by Paul U. Kellogg, known as the Pittsburgh Survey. This survey, mentioned in chapter 2, grew out of concern about the impact of the combination of industrialization and urbanization on the lives of the city's inhabitants. It was an ambitious attempt to describe a modern city using volumes of statistics, charts, and maps. But to Polansky, the usefulness of this type of enormous study is questionable. Following publication of the Pittsburgh Survey, he writes:

Pittsburgh, meanwhile, went on industrializing and urbanizing much as before; its fate was subject not to librarians but to the wishes of a number of supremely avaricious Americans whose descendants now control so much beyond Pittsburgh that an effort was permitted, a decade ago, to clean it up.

It is fair to say that the Pittsburgh Survey represents, for better and for worse, much of what social work research became during the next three decades. The study was truly monumental: imposing to look at, and perhaps a good place to hide the body, but not too useful. (1101–1102)

Polansky criticizes such "fact-gathering operations" passing as social work research and declares that a "total community inventory is a fatuous undertaking" (1102) on both theoretical and practical grounds. "Most community surveys were initiated in the hope that once the facts were compiled and artfully presented, the local leadership would be inspired to act." But, he concluded, "Experience in this country, alas, was that knowledge of needs did not guarantee incentive to meet them" (1102). Community surveys were too diffuse, ambitious, and superficial to promote change. Polansky was saying that social work research that wasn't used was of questionable value.

The second article in the 1971 *Encyclopedia* is clear about the benefits of research. Richard Stuart (1971) describes research relating to social casework and social group work. In contrast to prior entries to the *Encyclopedia*, Stuart's is combative in promoting the need for evaluative research in social work to root out "the last vestiges of the 'art versus science' controversy." His article provides a definition, description, and defense of experimental research in social work and introduces single-subject designs to the *Encyclopedia*. He acknowledges that, to date, "evaluative research typically yields such negative results that it is not especially useful to those agencies and individuals already committed to rendering service" (1107). Again, the use or nonuse of research is mentioned but not explored. The third entry that year, by Wyatt Jones, on research on social planning and community organization, is a two-page annotated bibliography of evaluations of organizations and community change.

By the 1977 edition of the *Encyclopedia*, research coverage had grown to five separate articles. The first, by Henry Maas (1977), is an overview of the organizational context for social services research in Great Britain, Denmark, and Canada and a smorgasbord of different types of studies, some classic, some recent, conducted by social workers and social scientists. It makes only passing comments about obstacles to collaboration among researchers and practitioners and about the use of research. This is followed by articles on interorganizational analysis (Morrissey and Jones 1977); the *Encyclopedia*'s first entry on program evaluation (Caro 1977), which notes that practitioners are often threatened by evaluation and may resist or discredit it; and an overlapping article on policy evaluation (Rossi 1977). The final article, by Polansky (1977), is an extended lament about how little effort the profession invests in research on social treatment. There "is no more than a handful of researchers studying social treatment at any one time," federal funding for research constrains creativity, anti-intellectualism is on the rise in society and the profession, and research on social treatment "has lost its momentum just at the point when it could have begun to achieve increasingly greater results" (1208). Although Polansky sees some signs of hope, he thinks research progress in conceptualizing practice theory has been meager. None of these five articles contains any specific discussion of the dissemination and utilization of research findings among practitioners.

In 1983, for the first time, NASW issued an interim supplement to the *Encyclopedia* to describe new developments and topics. Among the articles is one on "Research Developments" (Reid 1983), which notes that the locus for social work research is shifting from the agencies to the universities, where there is an increasing number of doctorally trained faculty; and that there has been a positive shift in the findings of recent studies on service effectiveness (131). He describes two emerging methodologies: developmental research and single-system designs (see the extended discussions of each of these in chapters 4 and 5). His discussion addresses the relevance of both developments for problems associated with practitioners' use of research findings. This extended discussion is the first time that the *Encyclopedia of Social Work* addresses the specific problems of research dissemination and utilization.

The subsequent editions of the *Encyclopedia* (the 18th, Minahan 1987, and the 19th, Edwards 1995), contain greatly expanded coverage of research topics. The former has eight research entries and the latter has ten; both include articles on developmental research, single-system design, research ethics, program evaluation, and measurement. Each subsequent volume recognizes problems of dissemination and utilization. Although by the 1980s the difficulties in the interface of research and practice were obvious, awareness of them had emerged slowly from attempts to ground practice more firmly in the activities of science. More easily acknowledged were earlier attempts—many only partially successful—to encourage practitioners to make better use of the research literature. It is to these efforts and what was learned from them that we now turn.

Assessing and Using Research

With the body of social science research relevant to social work practice and policy rapidly growing by the 1960s, some scholars urged that social work students be better equipped to assess and use research findings. This was relatively early in the development of social work research; the 1965 *Encyclopedia* contains one entry on the subject. In 1969, a University of Michigan faculty team of Tony Tripodi, Phillip Fellin, and Henry Meyer published a popular textbook specifically addressing this need, *The Assess-*

ment of Social Research: Guidelines for Use of Research in Social Work and Social Science, and an edited companion, *Exemplars of Social Research* (Tripodi et al. 1969a, 1969b). Because of the rapid growth of the social sciences and the development of social work research efforts, the authors note that practitioners need to evaluate critically and use this emerging knowledge, but most are not well trained either to produce or to consume empirical research. The standard research textbooks emphasize how to *conduct* research, but the authors identify the focus of their book as the

> *consumption* of research. Its aim is to increase the sophistication of the reader of research reported in the literature. The more one learns about doing research, the more sophisticated one is likely to become in reading research already done. Likewise, the more one becomes efficient in reading research, the more likely one is to understand how to do research. Nevertheless, there are different objectives and skills for the research producer and the research consumer. This book seeks to enhance the skills of research consumption. (1983:2, emphasis in original)

The book attempts to fill this need by providing pointers for classifying research reports according to type of research, offering key questions to ask in assessing studies, and suggesting a framework for using research findings in practice. The chapter on utilization of research focuses on the application of knowledge gained and offers guidelines for considering the utility of published research. Following the guidelines, the authors present a summary list of major questions the practitioner should ask of every article read. This exercise follows similar sets of questions used to classify the research and assess the adequacy of its methodology.

What is noteworthy about the authors' approach is the earnestness with which they present these complicated tasks. They assume that social work students need to be trained to consume, not produce, research. By providing multiple lists of questions to be used as handy aids, they encourage practitioners to routinely read, carefully assess, and then make appropriate use of the flood of empirical studies being published by psychologists, sociologists, political scientists, and social workers. The advice and guidelines are well intended and do target key considerations in the assessment of research reports, but the belief, some thirty years ago, that practitioners would have the interest, willingness, and capacity to perform this detailed and arduous process seems naïve today. Nevertheless, encouraging practitioners to read research studies and use them appeared at the time to be a logical and practical means of closing the acknowledged gap between the growing research enterprise and the world of practice. The structures and processes by which any profession transfers knowledge from researchers to practitioners is more elaborate, however, than critical reading skills.

KNOWLEDGE UTILIZATION

Social work scholars' attempts to prepare practitioners to consume research studies occurred against a backdrop of social scientists' and government agencies' concern about the dissemination and utilization of knowledge for program and policy decision making. Not only in social welfare but also in education, public health, social rehabilitation, community development, agriculture, medicine, and other fields, there was a stronger focus on how best to put knowledge to practical use. By the 1970s the application-oriented social science literature was alive with new ideas about social indicators, planned social change, technology transfer, and program and policy evaluations. An interdisciplinary effort in political science, economics, psychology, and sociology and in the professional fields of social welfare, education, and public health worked to elucidate the processes, structures, and impediments to the use of research-based knowledge. It was part of an attempt to bring science to bear on social policy and social practice, to rationalize public decision making and public social programs.

The question of how knowledge is used by practitioners and policy makers was important in the 1960s and received unprecedented attention from government agencies and scholars, most prominently at the National Institute of Mental Health in a program on knowledge transfer and at the University of Michigan in its Institute for Social Research's Center for Research on Utilization of Scientific Knowledge. But throughout the country, academic institutes and government agencies began working to improve the use of knowledge in public programs.

Researchers had been relatively oblivious to the processes by which knowledge, once developed, might be effectively disseminated and used. This is a complex phenomenon, as anyone would discover if they wandered into the exploding knowledge utilization literature of the 1960s and 1970s. The number of citations grew from 400 to an estimated 20,000 in two decades (Davis 1976). At the time, there were two compendia of that literature (Glaser et al. 1976, 1983; Havelock 1969). These volumes constitute massive summaries of the state of the art. For example, the 600-page *Putting Knowledge to Use* ends with a 167-page bibliography of relevant literature.

These books delineate the concepts, factors, phases, stages, structures, and models pertaining to the utilization of knowledge. They particularly emphasize the organizational factors related to knowledge utilization and change, using many schemes to summarize information. For example, many authors group together related variables that affect these processes. Other authors synthesize the literature by organizing it around stages in the knowledge-use process. Others attempted to build models to explicate how multiple factors interact at different stages of utilization. Three complementary and widely cited models were identified (Havelock 1969:11–15).

The Research, Development, and Diffusion Model (RD&R) suggests a linear, rational sequence that includes basic research, applied research, development and testing of prototypes, mass production and packaging, and finally, dissemination of knowledge. (See chapter 5 for how it influenced social work.) This model primarily pertains to the macrosystemic and policy levels of analyses. It assumes a planned and coordinated research and development process, a division of labor, and a separation of roles and functions. For example, the basic researchers are not the ones who mass produce, package, or apply the knowledge. In the RD&D scheme, consumers are a clearly defined and somewhat passive audience who will accept an innovation if it is delivered in the right way at the right time. Finally, the model stresses the scientific evaluation of every stage of the RD&D process. The high initial development costs prior to dissemination are expected to pay dividends in the long run in terms of efficiency, quality, and capacity to reach a large audience.

The Social Interaction Model (SI), unlike the RD&D model, assumes at the beginning that a specific diffusable innovation already exists and is relatively indifferent to the scientific or technical features that have gone into its development. SI is much more concerned with the diffusion of the innovation and how this is influenced by social structures and relationships. The model focuses on the microsystem level and on the latter end-use phases of the research, development, and diffusion process. It is more sensitive than the RD&D perspective to the complex and intricate set of human relationships that impinge on the diffusion process. It stresses the role of key influences in the end user's social network. Position in that network, face-to-face interpersonal contacts, and number of overlapping reference group identifications are all important in the diffusion process. The model is supported by numerous empirical studies in medicine, agriculture, and education.

In the Problem-Solver Model, the consumer of knowledge, the end user, is the center of analysis. This is basically a psychological and user-oriented model of dissemination and utilization. It assumes that self-initiated change offers the firmest motivational base and the best chance to achieve lasting effects. Change introduced by an outside agent may only be temporary. The model suggests that the problem-solving process begins with the consumer sensing and identifying a need, which is then transformed into a problem to be solved. A search is initiated to locate resources relevant to the problem and potentially feasible solutions are adapted to it. The solution is then used on a trial basis and evaluated in terms of need reduction. If the solution does not work, the cycle begins again. This general model informed the attempts of social work scholars to encourage practitioners to be critical consumers of research (discussed in this chapter) and to systematically evaluate their practice outcomes (discussed in chapter 4).

Each of these three models has strengths and limitations, and each found partial expression in efforts within social work to link knowledge production to knowledge use. The RD&D model emphasizes the role of research and development and the rational planning of diffusion efforts but pays little attention to the role of the consumer. A discussion of the uses of this approach in social work is in chapter 5. The SI perspective carefully documents the importance of social networks in understanding the flow of knowledge and utilization through a user system; however, it fails to articulate the linkages between the producers and users of knowledge. Only a few studies in social work have used this approach. The Problem-Solver Model directs attention to the psychological conditions under which new knowledge may be sought out and used by consumers but overemphasizes the extent to which consumers are able and willing to generate their own solutions to problems, although the use of single-subject research designs (discussed in chapters on empirical practice) can help them do so.

THE REALITY OF RESEARCH USE AMONG PRACTITIONERS

Since much of the knowledge utilization literature derived from studies of industry and agriculture, it was unknown whether it could be generalized to professional practitioners in social work or other fields. One of the questions in surveys of practitioners was to what extent they paid attention to research findings. Despite frequent claims that professionals use the latest knowledge, some early studies suggest that they often do not (Morrow-Bradley and Elliot 1986). A study in nursing illustrates the problem. Ketefian (1975) wanted to learn the extent to which a specific research finding was utilized by practicing nurses. The information, documented through a series of studies and widely disseminated over a five-year period, had a direct application to nursing practice. Nevertheless, the nurses surveyed were either totally unaware of the research literature or, if aware of it, were unable to relate to or use it. The nurses even continued to utilize methods they thought were unsound. Type and amount of education, recent graduation from school, and frequency with which they performed the procedure made no difference in the nurses' practice in this regard.

This problem was recognized in other professions as well. Accumulating evidence suggested that research findings designed to be relevant for professional practice were seldom utilized. This became a common frustration among those who conducted program and policy evaluations (e.g., Weiss 1972:10–11; Patton 1978), who discovered that their results seldom had any immediate impact on program or policy decision making. While social workers appeared to have a high regard for research, they didn't prefer their research courses as students, did not read much research-oriented material, seldom conducted or used research in their practice, and were suspicious of research findings that contradicted their beliefs about effectiveness (Kirk 1979).

By the 1970s, scholars were realizing that publishing a research report in a reputable journal did not constitute effective dissemination after all. Less than half of social workers read journals that contained substantial research content (Kirk, Osmalov, and Fischer 1976). (A recent survey by Mullen and Bacon [2000] suggests that the reading habits of social workers have not changed.) It was estimated that half the articles in each core scientific psychology journal are read by fewer than 200 persons (Garvey and Griffith 1967). Few mental health practitioners made any large-scale or systematic effort to read research results reported via formal communication channels (Roberts and Larsen 1971). And in nursing, medicine, and law, the major research journals were not widely distributed beyond the faculties of professional schools and full-time researchers.

Nevertheless, practitioners were exhorted repeatedly to make better use of knowledge. For example, Kamerman et al. (1973) prodded practitioners to use social science knowledge and provided a listing of theories, concepts, and disciplines with something instructive to say. Exactly how these ideas were to make their way into the heads and hearts of social workers is less clear. As discussed in chapter 5, perhaps the most systematic attempt to encourage practitioners to use social science knowledge was by Jack Rothman and colleagues at the University of Michigan, who culled the social science literature for empirical generalizations relevant to one method of practice, community organization. They formulated action guidelines and studied their efforts to have these guidelines used by practitioners and to disseminate the findings nationally. This unique, multiphased, long-term project to promote the utilization of social science research is described in a series of books (Rothman 1974; Rothman and Teresa 1978; Rothman et al. 1983).

The Problem of Research Use in Social Work

Since practitioners using research findings seemed to be a crucial, if imperfect, link between the worlds of research and practice, different hypotheses emerged about why this was not happening. One suggested that social science research was not immediately applicable to social intervention, either because it was too theoretical or esoteric or because it often focused on variables that couldn't be manipulated by practitioners, such as gender, social class, ethnicity, and marital status. Shirley Jenkins remarked, "The bottom line in research utilization in social work is that the research findings must have utility for practice. Elegance of design, sophistication of statistics, and soundness of theory are all beside the point if the problem that is addressed has little relevance to client needs or service delivery" (quoted in Rehr 1992:358).

Another common explanation was that researchers and practitioners had divergent orientations. They operated in different social worlds, were concerned with different types of questions, had different skills and, further, did not hold very high opinions of each other. Practitioners tended to rely on precedent, common sense, and intuition much more than on research findings. A third speculation was that since most practice-relevant research had not documented positive effects of intervention, it did not provide any useful prescriptions for practice. Finally, a broader problem was raised by an ardent defender of practice, Carol Meyer: "To affirm that there is more research is not to say that it is all of equal validity, nor is it to say that the field at large makes use of its findings" (1973:36). Further, she said, "Reliance upon empirical data has not been a hallmark of professional social work practice, partly because of our tools and objectives of research, but also, perhaps, because we have not yet agreed upon the goals and boundaries of social work practice" (38).

INTERPRETING THE PROBLEM

The 1960s witnessed the confluence of three emerging realities, of concern primarily to academics. First, the profession of social work was being called upon to be more scientifically grounded. Second, direct service practitioners, as the front-line workers, were the ones whose interventions needed to be informed by the latest research findings. Third, practitioners appeared to be disengaged or perhaps disinterested in utilizing research results.

With the difficulties of directly using much social science and the ambivalence toward research exhibited by some social work faculty and students, the gap between practice and research use appeared to be expanding at just the time when the profession was being challenged to be more research oriented and to scientifically justify its methods. The perceived need to produce more research and to use the results in practice had never been greater. And yet, antagonism was growing between those scholars who wanted to recast the profession in a more scientific mold and the vast majority of practitioners and some faculty who appeared reluctant to embrace science or its methods as their guide.

The fault line between research and practice took on a particular shape in the 1960s as academics struggled to understand the fracture between the production of research and its utilization by practitioners. This guided the profession for several decades. To appreciate how the antagonism was interpreted, we need to look at three seminal studies of practitioners' involvement in research for what they reveal about the making and shaping of the research utilization problem.

SYMBOLS AND SUBSTANCE

Careful examination of citations in the literature on the utilization of research in social work leads back to an early footnote to a study by Joseph Eaton, "Symbolic and Substantive Evaluative Research" (1962). This article became an obligatory source in nearly all subsequent publications. No one before Eaton had asked practitioners about their attitudes toward research, how they might use it, or how important they thought it was for themselves or their agencies. Although this seminal study is rarely analyzed in any detail in the text of later works, it achieved the status of a foundational footnote. Eaton's contribution shaped subsequent discussions, so a careful analysis of what he said will provide insight into the development of literature on research use in social work.

The article presents findings from a survey of 282 social workers in the Veterans Administration (VA) and a similar survey of nearly all the 4,000 employees of the California Department of Corrections. The respondents were asked to react to hypothetical situations, explaining what they thought their agency should do if it received a sizable amount of unrestricted new funds and how they would interpret and communicate to others several hypothetical research findings. Nine tables displayed the percentages of responses to items on the questionnaires. These data were presented simply and were easy to read, with no cross-tabulations, bivariate correlations, or tests of significance.

Eaton asked the VA social workers, "If your agency received a ten-thousand-dollar annually renewable gift, what would be your priority for its use?" The most frequent first priority (of 41 percent of the respondents) was to raise starting salaries to recruit more qualified caseworkers. The second most frequently cited priority (35 percent) was to "support work on a research project regarded as important by the agency" (424). When asked what they would do if they had the freedom to devote one day each week to some professional activity, the VA social workers responded that their first choice would be "do research." Their answers to several other questions expressed orientations supportive of research.

In addition, 159 mental health professionals in the California Department of Corrections were presented with two brief research summaries, one offering encouraging findings and the other offering discouraging ones. They were asked to indicate only one recommendation they would make for publicizing the findings. Their most frequent response to both summaries was to reexamine and continue the study to get more information. Their next-most-frequent responses were to publish it in a reputable journal and to issue the report to newspapers. Although there appears to be a slight tendency to publicize the encouraging findings more than the discouraging ones, Eaton does not indicate if the differences are statistically significant. Virtually no one recommended that the findings not be released.

The VA social workers were asked to indicate, for each of the four brief research findings, whether they would agree with, be doubtful about, or disagree with communicating the "facts" through each of five verbal and five written channels of communication. Eighty-six percent agreed that they would "report facts orally to [their] immediate supervisor"; 76 percent would "keep this fact to [themselves] until someone asks about it"; almost half would "report the facts orally to a special staff meeting of social workers, staff physicians and nurses"; one in five would "tell fellow workers [they] can trust"; and 15 percent would "informally tell [a] hospital administrator." One third would report the facts in writing to their supervisors and one third agreed that the facts should be included in the next monthly administrative report. Relatively few (16 percent) suggested publication in a journal, and almost no one suggested submission to a patients' newspaper or to the press.

Where Eaton went with these data could not be predicted easily from the data themselves. They could have been used to claim that social workers are supportive of research. His article's influence, however, came not primarily from his data but from his interpretation of them, which then echoed throughout the subsequent literature on the utilization of research.

From the beginning, the theme is the purity of evaluative research as science. The thesis appears in the second paragraph, not as a hypothesis to be tested but as a conclusion: "Research has two very different functions for organizations: a *symbolic* function and a *substantive* one" (422). Eaton leaves little doubt about the preferred kind of research—the substantive kind, free of bias and social constraints; symbolic research is tainted and receives only ritualistic avowal in committee meetings. Eaton claims that scientists pursue their work in an open-minded, even-handed, and objective way (426) and want their findings widely shared. He stresses the need for unhampered, open communication among scientists, distinguishing it from the secretive tendencies of business and governmental organizations (438).

A major message of his article, presented largely without data, is a description of the difficulty practitioners in large social service agencies have in using pure science. In the opening sentence of the abstract, Eaton pinpoints the problem as the contradiction between the avowed objectives of research and clinicians' acceptance of its findings (421). He claims that a favorable climate for research can be maintained only when it does not threaten those in power; therefore, organizations have a preference for symbolic rather than substantive research (422), because they control information for their own advantage. This theme saturates the article from its opening to its closing sentences, even though the data do not present such a clear, uncontested picture. In fact, little research of any kind is cited to support these assertions.

Another theme is that practitioners' fears about their bosses and their place in the agency lead to the corruption of science and the suppression of knowledge. Eaton works this theme from beginning to end. He begins with an astonishing statement—made without evidence—that in the professions, many of the findings of the more gifted practitioners are never written up or are filed away as inconclusive because of organizational concern for disturbing the status quo (422). From this unsubstantiated assertion about suppressed research, he moves with considerable liberty through his study's findings to conclude that practitioners have reason to react personally to data from evaluative research because the findings may raise questions about their competence. Research, he claims, is a personal threat to practitioners; only the strong and those with a "secure self-image" can handle it (427).

From these assumptions, he goes to great lengths to interpret his data as supporting the view that practitioners are reluctant to interpret research findings, particularly negative ones, and unwilling to communicate them. His own data, of course, do not seem to support this, so he interprets the respondents' apparently high regard for research as merely symbolic. In reviewing the data about their readiness to communicate findings, he argues that silence and ambivalence predominate, even though 86 percent of the social workers were willing to report research data to their immediate supervisors, a clearly appropriate channel. With written communication, he saw only great reluctance. Although it is plausible that verbal communications within organizations are the conventional way to share information, he jumps immediately into the morally loaded language of censorship, claiming that "academically trained members of professions exhibited self-censorship attitudes even when research findings had favorable implications" (436).

In his article, Eaton seems to tap latent suspicions among researchers that practitioners are not readily amenable to influence by scientific evidence. Ironically, his own data are largely irrelevant to his argument; he bases his conclusions on conventional wisdom, a scientific failing he generally attributes to practitioners.

RESEARCHER VERSUS PRACTITIONER

The 1960s was a watershed in the production of new social programs and, simultaneously, of serious efforts to evaluate them. Evaluation research, the attempt to use research methodology to assess the process and effectiveness of interventions, came of age. In 1967, *Social Work* published one of the first reports of these evaluative efforts to appear in a social work periodical. "Researcher Versus Practitioner: Problems in Social Action Research" (Aronson and Sherwood 1967) describes the efforts of two sociologists to evaluate antidelinquency demonstration programs in New York City, ground zero of the emerging War on Poverty. The authors mean to persuade readers of the central importance of evaluation and to describe problems that they encountered in performing it. The article contains no footnotes or citations to pertinent literature, which indicates how seminal it was.

The article points out that evaluative research was being required increasingly by funding agencies, in this case by the Ford Foundation and the President's Committee on Juvenile Delinquency and Youth Crime. A fortune was about to be spent on a variety of programs for delinquency and poverty, and without evaluation, no one would know whether the funds were well spent or wasted. The authors attempt to educate readers about the rational procedures that should guide the development of a demonstration project: setting objectives, specifying how they are to be achieved, and afterward, conducting an evaluation to determine whether they were. The important message is that for there to be an evaluation, the program must be conceptualized and specified in a way that appears rational to the researcher. The authors describe evaluation as a technically easy task, as long as the program planners and implementers do their job. The bulk of the article, as suggested by its title, is a description of how practitioners, by impeding the knowledge-development enterprise, make life frustrating for researchers.

In an anecdote, Aronson and Sherwood present the practitioners as myopic. The researchers were frustrated by their "preoccupation with the components of programs without reference to their objectives or to the connections between those components and the kinds of changes the program was intended to produce" (91). This preoccupation, they said, is illogical. They could barely get the practitioners to specify goals. The researchers complained that they had to do the conceptual work for the practitioners.

The authors describe the designers of the program as reluctant theorists and imply that the practitioners were slow learners who, for example, "frequently did not understand the necessity for the comparison group," "did not always keep to the requirements for maintaining equivalent groups," would not complete records, and sometimes were reluctant to provide clients' names (92–93). The practitioners were also fearful that the research would be used against them.

Difficulties were not limited to the research procedures themselves. The practitioners frequently would not implement the interventions according to plan: "Almost every social action program . . . was changed by the practitioners almost as soon as the program began" (93). Moreover, the authors recognize that program success often means something different to practitioners than to researchers. Commenting on this in relation to a summer-camp program, they present a condescending example of how a camp director *knew* that his program was a success because he could see its achievements "in the smiling faces of the campers," but to the researchers the frequency of smiling was not how the outcome variables were operationalized (94). Using their experiences, the authors continually contrast the rational requirements of scientific evaluation with the cautious, well-meaning, but narrow-minded preoccupation of practitioners.

We must remember that these were the early days of evaluation research, when researchers expected that rational decisions about social policy would be natural outgrowths of social experimentation. Social science research methodologies would be taken from the laboratory and applied directly to real-life programs. The practical and conceptual limits of experimental group designs were still not fully appreciated, and neither were the intricacies of practitioners' work. Consequently, the difficulties of evaluation research were blamed on those who were providing services but appeared less interested in or able to advance science.

Practitioner Views of Research. Aaron Rosenblatt (1968) was the first social worker to conduct a study of practitioners' use of research. It was reported in an influential article that appeared in a special issue of *Social Work*, when Scott Briar was editor-in-chief. Rosenblatt's data were from 308 practitioners in the New York City area who completed a questionnaire. They were asked first to think of a recent case in which they had encountered some difficulty in deciding on a treatment plan, and then to indicate which listed activities they would engage in before arriving at a decision, what they would do if they had more time, the value of different sources of help, the importance of different experiences that might have improved their practice, and which master's degree courses they found most helpful. The responses were reported separately, by rank order of what sources of help were valued more. Rosenblatt concluded that in each area, research was rated the least used or the least useful activity.

Rosenblatt's article is important because of the careful way in which the subject is made significant and interpreted. It builds on the concerns that had been so influentially articulated by Briar (1968a, 1968b), especially the need for social workers to develop and pay attention to an empirically derived knowledge base. In his opening two sentences, Rosenblatt states the case for the use of research in the profession. Social work research must be used to be of value; if it is not used, it is purposeless. Equally important, if it is not used by practitioners, researchers lose purpose. If what social work researchers have to say is not heard, or, worse, is not taken seriously and used, then the research enterprise is futile.

Finding that practitioners do not value research sufficiently, Rosenblatt offers three possible explanations. First, those who enter social work, through self-selection and the preferences of admission committees of graduate schools, do not have a scientific bent. Second, clinicians do not find research results especially valuable in solving practice problems. Rosenblatt, however, barely discusses these two possibilities. His third explanation, the one that he wants the reader to take the most seriously, is described in eleven paragraphs of reasoned speculation. It echoes Eaton: practitioners are unscientific in their thinking, protective of the status quo, and threatened by researchers, who, on the other hand, are the good soldiers of progress, championing objectivity and logic with the fervor of moral crusaders. While recognizing that the roles of researchers and clinicians should be complementary, Rosenblatt argues that the ideal relationship breaks down in reality. Practitioners find protection in fuzzy thinking that cannot be tested, making the profession vulnerable to fashion and fad.

Although he believes that the profession is impeded by the unscientific thinking of practitioners, Rosenblatt offers a sympathetic view of their predicament. Practitioners must retain faith in their approach to remain helpful to clients. For them to adopt the skeptical perspective of science would make them less effective. Moreover, researchers turn the tables, making practitioners, instead of their clients, the objects of study. This is likely to disturb the practitioners, making them uncomfortable and awkward. They recognize researchers' tacit promise of improving knowledge in the future as only a "promissory note to improve practice, which does not fall due until some indefinite future time" (58).

Rosenblatt concludes that for the sake of the profession, practitioners "need to make a long-range commitment to research," even though there may be few immediate payoffs. Clinicians should be altruistic for the benefit of researchers and the profession. He continues, "As part of that commitment they accept the need to support research, to co-operate with researchers, and to pay attention to research findings. They, not the researchers, will decide what to use and how it will be used" (59). He thus proposes an arranged marriage between practice and research in the hope that if they begin to live together, love will follow.

Framing the Research Use Problem. These three influential articles were thoughtful attempts to address perceived problems in the troubled interface of science and practice in social work. They were important not only because they were among the first during an era of concern about research utilization but also because they each went deliberately beyond their data to provide a broader analysis of the problems. In extending themselves, they were revealing a perspective common to a generation of researchers.

The shared themes can be culled from their explicit language and also from their latent messages. The first theme concerns the role of science and scientists. In each article, science and scientific research are presented as being of enormous value to the profession and to society. Science is the engine of progress, the source of light in a world of darkness, the carrier of reason into the realm of superstition. Scientists and researchers are disinterested observers of social work practice; their motives are innocent, their methods refined, and their hearts pure. Rationality and objectivity are the melodies of the researcher's song. Unfortunately, in the opinion of these authors, it is just not a song sung by practitioners with enough gusto.

A second theme concerns the impediments to science that are present in practice. Science has a rough time in the practice world because the forces of corruption are everywhere, setting traps for the naïve researcher. In each of these seminal articles, science is depicted as struggling against organizational banality. Social agencies are described as barely tolerant hosts of scientific work, providing temporary accommodations for the guest researchers but eager for them to depart. Agency administrators are interested only in findings that support the status quo or that place their agencies in a favorable light. Therefore, they may misinterpret, misuse, censor, or ignore the research that the scholars have painstakingly undertaken. In sum, these early authors warn, it is difficult to do science in the messy world of practice; it is much easier to do it in the quiet laboratory among fellow researchers with sympathetic views.

The third theme revolves around the assumed personal characteristics of practitioners, who, these authors maintain, are easily threatened. Practitioners are self-serving and self-protective and lack the intellectual power to be brave about unsupportive research findings. So they subvert research projects; use excuses such as the confidentiality of clients or commitments to service to be uncooperative, censor, or ignore unflattering findings; and, in general, are covert enemies of science. Moreover, the brightest undergraduates are attracted to other fields and the remnants who go to social work schools do not appreciate the significance of what researchers do. Practitioners are too human. Progress demands the dedication of the far-sighted and selfless.

Omitted from these portraits of research is any suggestion that researchers' motives may extend beyond the good and worthy; that scientists are not strangers to aggrandizement or status seeking; that the research process itself can be subjective and biased, sometimes fatally so; or that researchers may have a personal, as well as a professional, stake in persuading practitioners to value their work. There is little recognition that scientific technology has limits or that what researchers have labored to produce may not be particularly usable.

There is also little recognition in these articles that applied research makes new demands, not just on practitioners but also on researchers and their enterprise. Evaluative research, which is designed to be used, is more complex to implement than laboratory science because less is controllable. The methods of inquiry are more cumbersome, the data are more difficult to gather and less certain, and the statistical inferences are shakier. Not only organizational politics but also the natural turbulence of real-life practice impedes research. Furthermore, even in the sacred halls of science, discoveries often elicit skepticism or hostility, tolerance for unconventionality has its limits, and scientists themselves have no special immunity to bias.

Furthermore, there is not much in this literature of the 1960s about the practitioner as a dedicated pragmatist or the researcher as an unrealistic idealist. Practitioners may not have been ready to march to the researchers' drum, but solid knowledge about effective social intervention was not readily at hand. The research technologies for applied settings were not well developed. The small proportion of journal articles that reported research results were sometimes unreadable or trivial, and frequently both. Many articles were devoid of practical applications. Practitioners, whatever their unscientific misgivings about researchers, may have had justifiable reasons for ignoring them and their work.

The early studies of the use of research by Eaton, Aronson and Sherwood, and Rosenblatt raised a series of concerns and stimulated several efforts to improve the relationship between practitioners and researchers. It is noteworthy that not a single subsequent article on this topic was written by a practitioner; it has remained a preoccupation solely of academics. The further work consists of more studies on social workers' views and uses of research, how and what students are taught about research, and the many attempts to integrate the roles of researcher and practitioner (and the technologies of science and practice). But throughout these subsequent efforts are the background assumptions that research is not adequately appreciated or used by practitioners.

For example, a series of studies in the 1970s, some using national samples, raised similar concerns. Surveys of practitioners (Casselman 1972; Kirk and Fischer 1976; Kirk, Osmalov, and Fischer 1976) suggested that they don't like to study research, don't read many research articles, seldom use research studies in their professional work, don't do research in their agencies, and have difficulty accepting findings that challenge their beliefs. These conclusions were buttressed by two other small-scale surveys that suggested that social workers could not recognize common statistical symbols (Weed and Greenwald 1973; Witkin, Edleson, and Lindsey 1980). In yet another study, practitioners had less favorable views about the importance and usefulness of research than did graduate students in social work (Rosen and Mutschler 1982). In short, surveys of practitioners continued to paint a bleak picture of the status of research in the profession.

Recognizing the Many Methods of Research Use

By the 1970s, it had become customary to emphasize practitioners' nonutilization of research. That grim assessment had been based on a narrow view that expected those who designed programs and delivered services to read and use research results appearing in journals. Although some research use undoubtedly occurred in this way, the utilization process is far more varied, complex, and indirect. As some authors (Weiss and Bucavalas 1980:312) put it:

> Our understanding of research utilization has to go beyond the explicit adoption of research conclusions in discrete decisions to encompass the assimilation of social science information, generalizations, and ideas into agency perspectives as a basis for making sense of problems and providing strategies of action.

Dissatisfaction with the narrow view of utilization as direct application of findings to programs and practice has led researchers to different definitions (Pelz 1978; Rich 1977; Beyer and Trice 1982). They make a key distinction between "instrumental utilization"—using research for specific "decision-making or problem-solving purposes"—and "conceptual utilization"—allowing research to influence "thinking about an issue without putting information to any specific documentable use" (Rich 1977:200). Instrumental utilization preserves the classic notion of use. In conceptual utilization a user's decisions may draw on some combination of their own beliefs and research findings. This form can also be extended to use of terms, constructs, and other conceptual knowledge embedded in research reports (Tripodi, Fellin, and Meyer 1983:95). A third type of utilization, referred to by Leviton and Hughes (1981:528) as "persuasive," involves "drawing on . . . evidence to support a position." Advocates, lobbyists, policy makers, and agency executives who marshal scientific findings to promote a cause or a program are making persuasive use of research.

Two additional types are especially relevant to direct social work practice. One is methodological utilization, or use of research tools, such as single-case designs or standardized tests (Tripodi, Fellin, and Meyer 1983). A social work practitioner who employs the Generalized Contentment Scale (Hudson 1990) to trace changes in client depression over the course of treatment is engaged in methodological utilization. (The development and use of these techniques are discussed more fully in chapter 4.) The final category is indirect utilization, which, as we define it, involves use of theories, practice models, or procedures that are themselves products of research. The practitioner who employs a form of social skills training that was itself shaped by a research process is utilizing research, albeit indirectly. Unlike other forms, indirect utilization requires no direct exposure to research. The practitioner interfaces not with the research itself but with the practice approach based on it.

In social work practice, these different forms of utilization interact over time to produce effects that are difficult to trace precisely. The dynamics can be best understood through a systemic view of the direct practice community, which consists of lone practitioners, supervisors, program administrators, agency executives, researchers, educators, consultants, students, and others, who interact through publications, conferences, workshops, committee meetings, classes, consultations, and supervisory sessions.

This system draws on the broad domain of research and research-based practice methods in the social sciences and the helping professions. Although little is known about how it processes research products, we suggest that conceptual and indirect modes of utilization are the prevalent forms. Practitioners may seldom turn to research studies, but their practice may be more influenced by research than is commonly thought—through what they learned in graduate school, their use of empirical practice methods, books and articles that draw on research, program directors and supervisors who themselves are influenced by such literature, and data generated by their agencies. Thus, the research word is being spread, often indirectly and secondarily, and *in toto* may be affecting social work practice, although it is often difficult to observe. Moreover, such indirect utilization is vulnerable to distortions of the truth and self-serving sampling of findings. It is a complicated task to separate good from bad utilization, and few social work studies have attempted to do so.

We have not only overemphasized the instrumental form but also tended to overlook evidence of utilization, perhaps because so little has been found in studies of direct research consumption by social workers (Kirk 1979, 1990). Here we may be blinded by our own expectations. Hoping to find much more use than occurs, we have tended to equate little utilization with none at all. It may be more instructive to use a zero rate of use as our point of reference. In fact, this was probably close to the actual rate in the earlier years, when little research existed. Against this base rate, any evidence of direct use is a sign of progress. This prompts reinterpretation of some older studies, like those of Eaton (1962) or Rosenblatt (1968), which presented data that were generally interpreted to mean that social workers largely ignore research in their practice. Yet *some* of the respondents, on the order of 10 percent in the Rosenblatt study, were making use of research. Such small minorities should perhaps be identified and studied more intensively. There is ample evidence that instrumental utilization occurs often enough in the human services to justify this kind of knowledge-building effort (Glaser and Taylor 1973; Rich 1977; Leviton and Boruch 1980; Weiss and Bucavalas 1980; Beyer and Trice 1982; Velasquez, Kuechler, and White 1986).

Reid and Fortune (1992) have argued that research utilization of all kinds has, in fact, been increasing in social work, albeit slowly. Therefore, it is necessary to take a long view in order to discern the trends. They point to changes in education as well as in the agency environment as evidence. For example, there is little doubt that since the 1960s, graduate schools of social work have become more research oriented: more faculty have doctorates, more faculty are engaged in research, and it is more difficult to receive tenure without a record of research and publication. Furthermore, the research orientation has found its way into the curriculum and coursework. Courses in human behavior and practice are more likely to include material that is grounded in an empirical perspective. For example, in one widely used text (Hepworth, Rooney, and Larsen 1997), approaches such as relaxation training, cognitive restructuring, stress inoculation, and communication of empathy are included with citations of relevant research. Students' being taught to use these "research-hardened" technologies can be viewed as indirect utilization. This probably constitutes one of the major ways that research is transmitted to future practitioners.

Other efforts have increased use over the last quarter century. Schools of social work have replaced the requirement of a thesis or group project with requirements of additional courses that attempt to integrate research with practice, or with substantive fields of practice. Many schools have attempted to integrate practice and research teaching (Siegel 1985; Richey, Blythe, and Berlin 1987), introduce single-case evaluation methodology (Gingerich 1984), or have students use research to develop their own personal practice models (Mullen 1983).

The bottom line for research utilization is what happens in the field among practitioners. Probably the major impact of research on individualized services in social agencies is the same empirical practice movement that has so influenced social work education (Fischer 1981) (discussed in more depth in chapter 4). Its leading proponents have been academics who promote their views through writing, research, workshops, and teaching. The schools of social work, however, are only one medium for the spread of empirical approaches in agencies. Psychology, psychiatry, and education, among other disciplines, are influenced by behavioral and other forms of empirical practice. This has an important impact on service programs in various settings where social workers practice, especially mental health, child welfare, schools, and corrections. Thus, research-based interventions enter social work through many channels.

Although it is difficult to find examples of direct utilization in agencies of findings from a specific study of practice, there are instructive illustrations of how this may occur. For example, in the late 1960s Reid and Shyne published the results of a field experiment in which planned brief service did somewhat better than much longer extended service (Reid and Shyne 1969). The results were a surprise not only to the researchers but to the staff of the agency, which was a national leader in casework services and committed to a long-term treatment model. Planned brief treatment was only an ancillary service.

So what effect did the startling findings have on the agency itself? As the study appeared, the use of short-term treatment did increase, but the increase had begun *prior* to completion of the study and long before the findings were known. A few dramatic successes with clients at the beginning of the service phase of the experiment were made known in the agency by the project caseworkers. These successes supported the beliefs of key agency staff, who then encouraged greater use of short-term treatment.

From a researcher's point of view, the staff were jumping the gun in concluding that short-term service was effective before any systematic data were available (although the final results in this instance supported their judgment). This is an example of conceptual utilization. There was no systematic implementation of short-term service due to the final results, the response researchers traditionally look for. Instead, conceptual utilization led to gradual, spontaneous change, though it was technically premature. This may not be atypical of demonstration projects, in which staff may reach conclusions about success or failure well before the final results appear on paper or in journals.

Outside the agency, another nontraditional pattern of apparent utilization occurred. Executives or program managers who were favorably disposed toward short-term treatment invited Reid to give workshops at their agencies. The workshops, which were of course also favorable to short-term treatment, were then used to whip up enthusiasm for the approach, perhaps reflecting a blend of conceptual and persuasive utilization.

In still other instances, the Reid-Shyne study was directly influential on practitioners or agencies trying short-term service—examples of instrumental utilization. Even in these cases, however, one study probably was not the deciding factor. Rather, the practitioners and agencies were reacting to it in the context of an increasingly favorable climate of opinion about brief treatment, fostered not only by numerous other favorable studies but also by a burgeoning practice literature on short-term methods in social work and related fields. And, of course, the study, once published, found its way onto reading lists and started to be used in social work education.

The mixture of conceptual, persuasive, and instrumental utilization of short-term treatment research is, we think, typical of the way research influences practice. The original empirical base may be service experiments, demonstration projects, behavioral research, or some combination. It may be a springboard for opinion leaders' advocacy for new approaches through media such as articles, books, conference presentations, or agency in-service training. The innovations that prosper are usually those that strike decision makers as reasonable, needed, or exciting in the light of their own experiences or sources of information.

Examples of this process are a number of approaches in social work that have gained currency in recent years. In child welfare, tests of experimental programs in permanency planning (Emlen et al. 1978; Stein, Gambrill, and Wiltse 1978) and decision making (Stein and Rzepnicki 1984) led to large-scale training and dissemination efforts and to the development of field manuals for child welfare staff (Stein and Rzepnicki 1983). More recently, family preservation efforts have spread throughout the country, stimulated by demonstration projects (Fraser and Nelson 1997; Blythe and Patterson 1994).

In the fields of mental health, services to the elderly, substance abuse, and others, examples can be found of practice approaches that originated in research studies but made their way into specific program developments by means difficult to identify precisely. Instead, we see a complex force field of many factors that affect practice and practice innovations.

Improving Dissemination and Use

As reports about the problem of research use accumulated during the 1960s and 1970s, major social work organizations made efforts to confront the issues. In 1976, the Council on Social Work Education (CSWE) received NIMH funding for three years to analyze research utilization in social work, identify obstacles to it (especially those that existed in education), and recommend ways of achieving effective research use. CSWE sponsored several studies and one national and four regional conferences on these topics. The national conference was held in New Orleans in October 1977 and resulted in a monograph, *Sourcebook on Research Utilization* (Rubin and Rosenblatt 1979). A second volume published later, *Teaching Social Work Research*, contained papers on the different approaches to teaching research at the MSW level (Weinbach and Rubin 1980), and a third volume, *Research Utilization in Social Work Education*, contained final papers and recommendations (Briar, Weissman, and Rubin 1981).

During the same period, NIMH funded a parallel effort with NASW on the status of research in the profession. NASW organized a national conference on "The Future of Social Work Research" in San Antonio in October 1978 and published the major papers (Fanshel 1980). There was great attention to the growing importance of research to the profession, heightened awareness of what was commonly labeled the practitioner-researcher gap, and an emphasis on the need to improve the production and use of research.

Concerns about research utilization continued into the 1990s and were the topic of at least four separate major efforts. The first was a May 1989 conference sponsored by Boysville in Michigan that resulted in a book, *Research Utilization in the Social Services: Innovations for Practice and*

Administration (Grasso and Epstein 1992). A second effort, funded by NIMH, was the Task Force on Social Work Research, which issued a major report in 1991, *Building Social Work Knowledge for Effective Services and Policies: A Plan for Research Development.* A third national conference, sponsored by the Columbia University School of Social Work in March 1993, resulted in *Practitioner-Researcher Partnerships: Building Knowledge from, in and for Practice* (Hess and Mullen 1995). Finally, the George Warren Brown School of Social Work at Washington University hosted a conference on developing research-based guidelines for practitioners in May 2000 and published *Developing Practice Guidelines for Social Work Intervention: Issues, Methods, and Research Agenda* (tentative title; Proctor and Rosen in press).

All these efforts show concern about the "gap" between research and practice and make suggestions about how it can be lessened. But the discussions no longer consider that the answer is simply to entice practitioners to become better consumers and users of research articles. There are many other proposed ways that the gap can be bridged (some explored in other chapters), such as producing more relevant research, training scientist-practitioners, developing scientifically based guidelines, and creating mechanisms for researcher-practitioner collaboration. It is a sign of progress that we no longer have observers wishing that social workers would cultivate better reading habits. Nevertheless, students' and practitioners' use of research has remained a concern, clearly embedded in the most recent comprehensive review of the role of research in the profession, conducted by a national task force appointed by the Director of NIMH (1991).

A long and broad view of research utilization in social work practice, from the surveys of Karpf to the present, shows that progress has been made on many fronts. What is more, both graduate schools' and agencies' infrastructures are developing mechanisms that should accelerate progress in the next decades. However, much more needs to be done before ordinary social work practice can be considered scientifically based in any real sense. Among the possible directions to pursue, two appear particularly promising: the generation of useful practice technologies and the development and application of methods of diffusion and implementation.

Generating Practice Technologies. As we have argued, advances in indirect and methodological research utilization in recent years have occurred in large measure through the creation of research-based technologies, including practice approaches, methods of single-case study, information systems, and computer applications. However, the methods must be useful, and by that we mean defined as useful by practitioners themselves. Ashford and LeCroy (1991) suggest that the problem of research utilization in social work is that the questions practitioners ask are not directly addressed by researchers' answers. They make a plea for more attention to how social workers use knowledge in their work, what level of knowledge they use, and how they make decisions in practice. This may account for why researchers are often surprised, if not disappointed, by practitioners' reactions to their work. As one sociologist (Eriksson 1990:15) recounted, following a research project that was designed to be useful to practitioners working with drug abusers:

> The reaction during and after the research period, when reports and books were written and presented and speeches given, produced a big *surprise*: What I had thought was useful, the audience and the readers didn't consider to be useful at all; what I had thought was useless for social work and only of interest to theoretical sociologists was judged as useful.

The continued development and testing of practice models is central to generating useful methods and may well be the single most important contribution that research can make to practice, just as the development of effective medications has been a boon to physicians. But it is not enough for researchers to produce practice models. The models need to be tested, evaluated, and revised with the kind of practitioners and clients for which they are intended (see chapter 5). Adaptations for particular settings need to be tried out. Practice protocols need to be sufficiently detailed, comprehensible, and flexible to provide realistic guidance for those who use them. Models shaped by this process have attained what Thomas (1985) has called developmental validity, a critical requirement for a useful practice model.

The same thinking applies to methodological utilization. Gingerich (1990b) has argued for the need to assess single-case methodology to discover if, and how, it may be useful to practitioners. More generally, the notion of treatment utility (Hayes, Nelson, and Jarrett 1987) can be applied to any type of methodological utilization. For example, what benefits do practitioners perceive from using Hudson's (1990) computerized clinical assessment package or from data on client change from the information system at Boysville (Grasso and Epstein 1987)? What difference does this information make in terms of intervention processes and outcomes? The primary purpose of such studies would not be to demonstrate how valuable all this methodology is in practice, although it is hoped that some benefits would be documented. Rather, the studies would be designed to discover how the utility of this methodology could be improved.

Diffusion and Implementation. No matter how potentially useful they may be, neither practice methods nor research findings will be adequately employed without better methods of diffusion and implementation. This also applies to the dissemination of practice guidelines (see Howard and Jenson in press; Videka-Sherman 2000; Mullen and Bacon 2000). We need to find more effective ways to acquaint practitioners with research products (diffusion) and get personnel to actually use them (implementation) (Robinson, Bronson, and Blythe 1988). Traditionally, a principal means of diffusion has been published articles, which, as has been well documented, practitioners pass over (Kirk 1990).

Perhaps practitioners would be more inclined to read research if the articles were written for them. Researchers may think they are writing for practitioners if they publish in journals such as *Social Work* or *Families in Society* (formerly *Social Casework*) and if they go light on methodology and stress implications. But, as Schilling, Schinke, and Gilchrist (1985) argue, reaching practitioners requires a radically different format that emphasizes practice methods or knowledge rather than research methodology. Furthermore, practitioner-oriented articles that summarize and interpret existing research in particular areas are needed. Unlike conventional reviews, these articles would feature useful knowledge and methods and illustrate potential applications. In general, research-based information should be presented boldly, as Schilling et al. (1985) observe. In research terms, the primary concern should be to avoid withholding possible useful information (Type II error) rather than making interpretations that might be wrong (Type I error). Obviously the two types of errors must be balanced, but in the practice arena, where decisions are often made on the basis of a metaphorical flip of a coin, a metaphorical Type I error rate of 20 or 30 percent may be worth the risk if the margin of error is made clear (Fortune and Reid 1999).

The academic establishment needs to provide more, and more sophisticated, initiatives for implementation of research-based technology. One-shot workshops or consultations are inadequate. In-service training projects in which staff help shape the technology to their own purposes and trainees are provided feedback about actual case applications are required. Such projects are reported by Kazi, Mantysaari, and Rostila (1997), Kazi and Wilson (1996a, 1996b), Mutschler (1984), Rothman (1980), Rooney (1988), and Toseland and Reid (1985). Training and adaptations should be informed by the literature on utilization. Both agency and academic establishments must provide better rewards for these activities, otherwise few will be undertaken.

Traditional reviews of research use are discouraging and will continue to be so because they neglect the realities of how knowledge—from research or other sources—is incorporated into what social workers and social agencies actually do. New ways of thinking about and studying research utilization processes in social work practice are needed.

Conclusion

Ever since the organizational and intellectual divorce of social reform from the social sciences, those concerned about the development of social work practice have recognized that the profession must maintain some working relationship with the scientific enterprise. Social work could not, with impunity, walk away from science. The dominant view from the 1930s until the 1980s was that this relationship could be maintained by having practitioners absorb directly the fruits of an expanding social science research effort without having to associate with social scientists. Practitioners were encourage to read new findings critically and to apply the kernels of truth they gleaned to their work with clients in whatever way they saw fit. This strategy depended on professional journals to serve a mediating role, allowing scientists to wash their hands of the dissemination business as soon as their articles hit print and practitioners to feel connected scientifically by the arrival of each month's issues. Journals provided the illusion for both scholars and practitioners that research was being produced, disseminated, and used, without the two groups directly interacting.

As soon as scholars started asking practitioners about their reading habits, their research knowledge, and their attitudes toward research, the illusion shattered. The worlds of scientific inquiry—populated by social scientists and the growing ranks of social work researchers—and social work practice were deeply estranged. At first, there was a strong tendency to interpret the fracture as caused by practitioners' failure to value, read, and consume research sufficiently, but soon it was recognized that there were failings on both sides, that researchers' studies were often flawed, of limited generalizability, irrelevant to the concerns of practitioners, or simply indigestible. Furthermore, research use—the link that would bind these separate worlds together—was a complicated and multilayered phenomenon that might occur through indirect methods of which neither researchers nor practitioners were fully cognizant. The topics of how practitioners learn and how study findings make their way, eventually, into use in social work are relatively unexplored but important questions for the future.

CHAPTER NINE

Knowledge for Practice: Issues and Challenges

S ince Flexner laid those important challenges before the audience in Baltimore in 1915, the profession of social work has struggled continuously to join social researchers and social work practitioners in fruitful collaboration. These attempts stem both from the need to have social work recognized publicly as a credible profession and from the belief that services will be more effective and clients better served if the methods of practice rest on sound foundations.

All professions strive for that legitimacy, but they differ in the extensiveness of their knowledge base, the character of the inquiry that produces it, and how they use it in practice. Law and medicine, for example, use different methods to generate knowledge and have different mechanisms for connecting that knowledge to practice (Staller and Kirk 1998). Legal knowledge consists of principles derived from legal precedents; legal research consists of sifting through case decisions in search of principles that can be used to a client's advantage in a public adversarial process. Practice and research in medicine conform more closely to social workers' ideal for their own knowledge: a carefully developed body of theory and facts, validated by objective, controlled clinical trials, is applied by skilled practitioners using sophisticated technology. This is still more of an aspiration than a reality in social work.

This book has described and evaluated some of the major efforts that social work has made to forge links between science and practice, ideas and application, theory and activity. Now at the end, it is only proper to reflect on the progress made in developing knowledge and to consider the enduring challenges.

Sharing Knowledge

Private knowledge—known only to a specific person—is of no particular note for any discipline or profession. Whatever its validity or potential usefulness, it will not enter into the accumulated wisdom of the field unless it is shared with others. Moreover, its validity and generalizability are likely to be established only after it is communicated to others who can refine, test, or employ it in some way. Written expressions of scientific knowledge are merely claims by the author that must be reviewed for their evidentiary basis and generalizability, initially by editors and expert reviewers and later by readers, who may subject the claims to further tests involving reanalysis of data, logical integrity, or replication. Through these processes, claims are verified, modified, or rejected. Over time, a shared knowledge base is developed and disseminated.

Practitioners' participation in these processes is limited. They may collaborate in the development of innovative interventions and implement them in experimental tests. However, building knowledge tends to fall into the domains of academicians and researchers. Practitioners are certainly involved in other kinds of information sharing and other modes of communication within the social work community. Colleagues share observations, experiences, and opinions with each other at conferences, in staffing meetings, and over coffee. The information could be about practice techniques or client problems and could be both useful and valid. It could be directly or indirectly derived from the published scientific literature or preliminary ideas that will never be developed into contributions to the published body of knowledge. But as long as the observations are communicated only informally, verbally, and among a few colleagues, they remain apart from the profession's established knowledge. Adding to the knowledge base involves making thoughtful written contributions to the literature.

Getting into print is not easy for practitioners or academics. Even after research has been completed, preparing a report for publication—reviewing the literature, presenting the problem, describing the methodology and findings in detail, drawing conclusions, and expressing them in prose that is succinct and easy to read—takes considerable time. Few articles are the result of a single draft. Experienced authors know that they will revise manuscripts many times before submitting them to journals, and will revise at least once during the peer review and editorial process. Writing takes time and devotion, usually lots of both.

Even when the manuscript is finished and sent to an appropriate journal, acceptance is far from assured. *Social Work*, the profession's major journal, accepts only about 10 percent of the manuscripts submitted for consideration. Other social work journals have somewhat higher rates of acceptance, but few publish even half of what is received. Thus, taking the time to write a research report is more likely to result in an unpublished than a published piece. This explains why most published articles in social work are written by professors or others affiliated with academic or research institutions. They have advanced training in research and scholarship, the time to write, and a variety of incentives to publish, including promotion, tenure, and salary increases. Even so, only a small minority of faculty in social work (and in many other fields) contribute a majority of the published journal literature. Publishing takes enormous effort; even among those trained and encouraged to do it, relatively few succeed.

That is why very few social work practitioners are actively involved in research or writing for publication. They are not hired or paid to do research or publish; few have the necessary skills; and most hardly have the time to spare. Nor do they have the institutional incentives to do formal research and publish the results. Thus, knowledge production and dissemination is largely in the hands of a small number of academics, while expectations for use have often been placed on the vast number of practitioners.

Recent Reaffirmation of Old Concerns

Ever since the profession of social work divorced itself organizationally from the social sciences, there has been a need to anchor practice and practitioners to a knowledge base. But in each era of its development, the profession has been confronted with reports that questioned how effectively this was being accomplished. Several recent reports chart the limited progress in bringing research-based knowledge into social work practice. In 1988, the Director of the National Institute of Mental Health appointed thirteen distinguished social work leaders to a Task Force, chaired by Professor David Austin of the University of Texas, to conduct the most comprehensive analysis and examination of the knowledge base in the social work field to date. It gathered and analyzed data from twenty-one different studies; reviewed the research content of journal articles, course syllabi, and common practice textbooks; reviewed the curricula vitae of faculty; surveyed faculty, doctoral students, deans, practicum directors, and NASW members; and examined the sources and patterns of research funding. After three years, the Task Force issued a report (1991) in which it concluded that too little social work research was being conducted. Further, what was being done was inadequate to meet the profession's need for knowledge or society's need to address increasingly complex social problems. Compared to other professions, social work lagged far behind in the development of both basic and applied research. In a nutshell, said the Task Force, "there is today a crisis in social work research" (11), of which it identified eight aspects:

- There is a paucity of researchers. In a profession of more than 400,000 members, few conduct research and even fewer publish anything that leads to the development of a cumulative knowledge base. (Fewer than 900 have published any research at all since 1985).
- Those who do publish, publish very little, and it demonstrates limited research skills.
- A critical gap exists between the knowledge needs of practitioners and the foci of researchers.
- Dissemination of research is fragmented and inadequate.
- At all levels of social work education, the teaching of research skills is inadequate.
- At the doctoral level, training for research careers is uneven, inadequate, inefficient, and inaccessible for many.
- Organizational and funding support for research development in social work are grossly insufficient.
- Few social work researchers are included in national bodies that determine research priorities, policies, and funding decisions.

(12–15)

Karpf's concerns from 1931 echo throughout this 1991 report. The Task Force concluded that institutional supports for research in social work are severely limited. Further, none of the major organizations or institutions, such as NASW, CSWE, or NIMH, has offices to promote social work research, and "few social work researchers are included in the national bodies that determine research priorities and government research policies pertinent to social work practice" (14–15). The Task Force also noted, as had earlier surveys, that little of the research that is published is disseminated and applied in practice. Most practitioners read few journals; no more than 5 percent of NASW members read social work journals on a regular basis.

Seven years later, Austin wrote a follow-up report of progress made by the profession (Austin 1999). On some fronts, there was encouraging news. NIMH was pumping more money into social work research, seven new social work research centers had been established, there were more research training workshops and technical assistance workshops, national social work organizations were more concerned about the infrastructure for research, and a new research journal had been started. On other fronts, there was continuing concern. Over two decades, despite the growth in the number of doctoral programs in social work, the number of Ph.D. graduates per year stayed the same; only 60 percent of students admitted to doctoral programs appeared to graduate, and many of them were not headed into research careers. Finally, it was unclear to Austin, as previously to others, whether and how any of the increased research activity among academics was making its way into the practice community.

The usability of research by practitioners is the focus of a recent review of the content of articles in social work journals (Rosen, Proctor, and Staudt 1999). The reviewers examined all 863 articles in 13 journals from 1993 to 1997, noting whether they presented research and, if so, categorizing it. Assuming that outcome studies of social work intervention were the most usable type of knowledge for practitioners, the authors found that only 47 percent of the articles contained reports of original research; only 15 percent on intervention research specifically. Of those, less than half described interventions in enough detail to permit replication by practitioners or other researchers. The authors' conclusion was that the most important type of research that could guide social work practitioners, i.e., intervention research, was rarely conducted. Therefore, should practitioners venture into the world of research articles, they will not find much of use. In short, the research that is conducted is often not directly helpful to practitioners.

The author of a 1994 study presents a parallel finding. Rosen (1994) asked 73 experienced practitioners to provide rationales for the treatment-related decisions they made with a total of 151 of their clients. The workers were asked to complete forms explaining their rationales; these narratives were coded into seven categories (e.g., theoretical, policy, values, client's wish, instrumentality, empirical evidence, personal experience). "Empirical evidence" was liberally defined as "any allusion to a research study, a research-based generalization, naming a study or author, or simply stating 'research has shown'" (573). Nevertheless, of the 771 different rationales gathered, only 2 were coded as alluding to empirical evidence. Rosen concludes that "The lack of use of research knowledge as a basis for making practice decisions challenges social work's basic professional aspirations" (574).

This result is clearly disappointing. Indeed, few if any studies of practitioners show that they link their treatment decisions explicitly to the scientific literature. This failure has many causes. First, the literature is itself not well organized. Even traditional summary reviews are problematic. They are often unsystematic in what they include and how they evaluate it, so many literature review articles are subjective, scientifically unsound, and inefficient as a method of summarizing knowledge (Light and Pillemer 1984). Second, journals usually publish the results of single studies, which are rarely sufficient grounds for interventions because of their limited generalizability. What is reported in any article should always be considered tentative until there is replication by others in different settings, using other populations, as discussed in chapter 7. Only in this way will findings be refined and qualified properly and their specific niche in the body of knowledge found. This may take years. Under these circumstances, it is reasonable for practitioners not to alter their practice on the basis of one or two articles.

Third, social work knowledge is scattered across more than thirty fields and among micro, mezzo, and macro levels of practice. There is considerable knowledge within the social work profession, but it is not necessarily linked in any systematic way, nor does it necessarily develop evenly or through any established, sanctioned process. Tucker (1996) argues that this lack of consensus about an organizing paradigm in social work has impeded its knowledge development and professional status. The fragmentation explains why one observer could describe the social work literature as embarrassingly rich (Kadushin 1959) and a few years later conclude that "Social work has not produced a systematic body of knowledge" (Barlett, Gordon, and Kadushin 1964:iii).

There is a lack of structure guiding the acquisition and building of social work knowledge (Staller and Kirk 1998). Social work knowledge comes from diverse sources, not only from the social sciences, governmental policy and guidelines, and scholars who produce theory and research but also from practice wisdom (see Klein and Bloom 1995; Zeira and Rosen 2000). Each of these sources does not contribute evenly to all areas of social work, and there are no established rules for determining what knowledge should be incorporated into the profession and what excluded. For example, on what grounds or through what processes should social work incorporate the concepts of the underclass, codependency, repressed memories, or the battered child syndrome? There is limited consensus on the methods for establishing valid findings or determining what kind of systematic inquiry is preferable for knowledge building. For example, social work has no guiding commitment to precedent (as does law), clinical field trials (as does medicine), or established dogma (as does some religious leadership). Moreover, there is limited agreement not only on how and what to add to the knowledge base but also on what constitutes the current base on which to build. In sum, social work has knowledge, but it is difficult to describe, add to systematically, or identify clearly when used by practitioners.

It is indeed reasonable to hope that the profession of social work will develop a knowledge base grounded as much as possible in many types of scientific inquiry. But because of the breadth and complexity of social work practice, it is unlikely that the foundation will ever be easily delineated or summarized, or even be discipline-specific (see Thyer 2000), since so much of it is scattered across many disciplines. Consequently, mechanisms of developing, accumulating, and disseminating knowledge to practitioners will be equally scattered, complex, and difficult to trace. This constitutes not a failure of research in social work as much as an inevitability.

Signs of Progress

This conclusion appears overly critical, as if little progress has been made or as if we know little more than previous generations. Progress may have been slow, but in the past half century, the research base relevant to social work practice has expanded enormously. For example, researchers in the helping professions have conducted literally thousands of controlled evaluations of interventions that are being used, or could be used, by social workers. Knowledge relating to understanding and assessing clients is being produced at an increasing rate by both the helping professions and the social sciences. Unlike practice schools and movements that wax and wane, the social work-relevant research has been marked by slow but steady growth since the beginning of the profession.

As we have suggested above, this knowledge is probably not being used *directly* to any great extent by social workers in their practice. That is, only rarely will practitioners link their activities to particular studies. But as we have argued in chapters 7 and 8, practice has begun to be affected by utilization of research products, such as empirically tested intervention programs. Practitioners who use research-validated methods of exposure therapy or group interventions with caretakers of the frail elderly without specific knowledge of their empirical credentials are nonetheless making use of research. Conceptual utilization seems also to be playing a role (chapter 8). Practitioners may be influenced by general knowledge of the research support for a method, combined with their own experience in using it successfully. In a recent survey of clinicians in a large agency, over 40 percent of the 64 social workers responding indicated that "research findings had changed their practice" and were able to give specific examples of methods they had adopted because of perceived research support (Mullen and Bacon 2000). As practice-relevant, research-based knowledge becomes more plentiful, it is more likely to be used, especially when promoted by research-oriented faculties in schools of social work, government agencies, managed care organizations, and practice guidelines.

Although the use of scientific methods in practice, epitomized in recent years by the single-system design, has not progressed in the way its advocates had hoped, there is still evidence that an imprint has been made. It can be seen, for example, in the small minority of practitioners who make some use of single-system designs (chapter 4) and in the larger, and perhaps increasing, proportions that use standardized instruments. In one of the two most recent surveys that we were able to locate (Marino, Green, and Young 1998), 60 percent of the practitioners reported use of such instruments; in the other (Mullen and Bacon 2000), the comparable figure was 30 percent. Both studies report higher usage than documented in earlier surveys.

There is also reason to believe that social work agencies collectively are becoming more involved in research and research tools. The growing use of computers and management information systems are a part of this trend. Child welfare practitioners using the federally mandated SACWIS information system now operational in most states (chapter 6) can conduct "instant studies" at their desktop computers, which might inform them, for instance, of how long different minority groups in their regions remain in foster care. There is also evidence of appreciable practitioner involvement in agency studies. In one recent survey of graduates of a school of social work, almost half reported participating in agency-based studies, principally needs assessments and client satisfaction surveys (Marino, Green, and Young 1998).

These "signs of progress" can be dismissed, of course, as mere ripples in a practice world controlled far more by ideology and convention than by scientific knowledge and method. Still, in the long view they represent significant steps forward—from a baseline of practically zero a half century ago. Even though scientific knowledge and methods may currently affect only a small fraction of practice activities, they have begun to make a difference. And perhaps here and there the difference may be critical.

Common Challenges in Applying Knowledge to Practice

Despite the gains made, the application of social work knowledge in practice has often been impeded. One factor is that practitioners and researchers occupy different roles in different institutions. As we have discussed in many chapters, it has been academics who have tried to convey the methods and findings of science to the practice community. University faculty, by role and disposition, are more comfortable with the abstract, the theoretical, and the general case than with the peculiarities of specific practice realities. They are also trained skeptics, who are often more comfortable questioning each other's and the practice community's conventional wisdom than offering realistic alternatives. The social work profession's academicians are relatively few in number, and even if most of them embraced a particular perspective, which they rarely do, their views would not immediately have much impact on practice. Researchers have few means of speaking directly or effectively to practitioners, and the distance between them is exacerbated by the lack of any established mechanisms for mediation or translation. How do theoretical ideas get translated into practice techniques and used on a trial basis? How do practitioners suggest problems and feedback to researchers? What institutional structures encourage such collaboration?

PUBLISH AND RUN

For decades this absence of effective mediation could be disregarded because researchers assumed that dropping a research report into a professional journal and running off to write their next manuscript was an effective dissemination strategy (see chapter 8). In this scenario, practitioners were expected to scan the professional journals each month looking for research findings that would inform their practice. The journals were viewed as the means of speaking directly with practitioners (and, of course, with other researchers) and of providing knowledge-based guidance to the field. This approach overestimated practitioners' capacity and motivation to consume research and use it effectively. But it also often overestimated the scientific and practical value of the literature.

OVERESTIMATING USEFULNESS OF SCIENTIFIC METHODS

As we reviewed in the first three chapters, the scientific foundation for social work practice remained shaky for many decades, perhaps well into the mid-twentieth century. Scientific charity didn't contain much science; nor did early casework, psychoanalysis, psychiatry, or efforts at social reform. This was due in part to the fact that scientific methods as used in applied social science were still being developed and refined. As we described in chapters 2 and 3, the research methodologies were crude, by today's standards. There were few attempts to systematically study the effectiveness of intervention at the client, program, community, or policy level. Thus, in retrospect, what knowledge researchers or journals had to offer to practitioners was of limited value. What was arguably of more value was the scientific habit of mind—of gathering information systematically about a case, considering alternative explanations for the client's difficulty, using intervention deliberately, and following up on the case. This style of rational problem solving, borrowed from science, has remained an aspiration for both social work researchers and practitioners.

ABSENCE OF INFRASTRUCTURE

As the profession developed a greater capacity to generate research-based knowledge in the 1960s, when doctoral programs and social work education expanded, the evolving research enterprise grew fragmented and disconnected from important professional structures that could nurture it. As we have described, the knowledge base for social work practice is spread across types of client problems (diagnosis), levels of intervention (individual, family, group, etc.), fields of practice (health, child welfare, etc.) and other dimensions (theoretical viewpoint). There has been no consensus regarding how the knowledge should be organized or whether there should be clear priorities determining which research problems should be addressed. Moreover, the research enterprise in social work is largely individualistic. There is no shared master strategy for developing the practice knowledge base; researchers each pursue their own interests, define their own topics, select their own methodological approaches, and work at their own pace, usually somewhat isolated from most of their colleagues and from the practice community. There has been no designation such as the "Decade of the Brain" announced by NIMH or coordinated efforts such as in the biomedical field to study AIDS, cancer, or heart disease, where programs of sustained federal funding guide the research efforts of board communities of researchers.

In fact, the federal government has never made social work research a priority, and only recently has NIMH targeted some funds to social work researchers—and then only to work on identified mental health issues. Certainly, other agencies at the federal and state level have funded social work researchers, but usually in response to some particular policy concern—to move more women off welfare or to increase the rate of child adoption—rather than to develop knowledge to improve social work practice in general. There is nothing wrong with this pattern of public funding; some of it does support social work researchers. It does not, however, provide for any coherent identification of research problems that might have wide applicability for social work intervention.

The diverse research efforts within the social work academy face another shortcoming, namely, that the leading professional social work organizations have historically had a weak, largely symbolic interest in research, e.g., sponsoring a conference every decade or so. In fact, many researchers view both the Council on Social Work Education (which accredits BSW and MSW programs) and the National Association of Social Workers as lost opportunities. These organizations have been much more concerned with public relations functions and their relationship to their broad, largely non-research oriented constituencies than with the need to nurture and coordinate social work's research efforts. In fact, over the last decade, out of some frustration with these established organizations, groups of leading researchers launched the Society for Social Work and Research; incorporated an existing journal, *Research on Social Work Practice*; and created a new annual research conference. In short, social work research has been impeded in its development and dissemination by the absence of powerful, supportive professional organizations that would link scattered researchers to the practice community.

GREAT EXPECTATIONS FOR PRACTITIONERS

In the absence of coordinated strategies of linking research to practice, and in the face of evidence that practitioners don't use research, some academics have tried to disseminate scientific technology, rather than research findings, to the field. These attempts to use science as a method of practice found expression in the efforts to disseminate single-subject designs (SSD) (chapter 4) and computer applications (chapter 6) to practitioners. In each case, new technologies that were enormously useful to the research community were exported to social work agencies. Instead of being required to read and consume research findings and modify their practice accordingly, practitioners were asked to use these tools of science to increase the rationality of social work. For example, SSD asked practitioners to work more systematically, measuring client problems, intervening at specified moments, documenting outcomes, and analyzing and interpreting data on the effectiveness of their interventions. This was an attempt not just to place research close to the practitioner but to make the practitioner more research-minded.

Early advocates of the use of computers also wanted practitioners to use new tools: management information systems, decision support systems, expert systems, clinical information systems, etc. This involved not just new ways of thinking but a new reliance on using electronic devices wired to large databases. While the SSD techniques required restructuring the timing of some activities with clients, computer-assisted practice inserted machinery in the interaction between the worker and the client. Skeptics worried about threats to confidentiality and privacy, and as with SSD, critics were concerned about the dehumanization of an essential interpersonal relationship.

The advocates had great expectations that, with some tutoring and encouragement, practitioners would adopt these technologies rapidly and easily. This did not transpire to any significant degree, except in a paradoxical way. Although single-subject designs were not adopted explicitly by practitioners as tools for self- and client improvement, the trappings of the method are now commonplace in the world of managed care—but they serve cost-controlling, not knowledge-developing, functions. Practitioners frequently find that they are reporting to health and behavioral science managed care companies about their clients' specific problems and level of functioning, indicating courses of particular modes of time-limited treatment, establishing benchmarks for improvement, and carefully monitoring progress. Similarly, computer technology is now common in social agencies, used primarily for administrative accounting purposes and for communication among staff. While these are undoubtedly useful functions, they hardly yet constitute practitioner reliance on data and probability equations to make clinical decisions. Nor are these the core functions promoted by early proponents of computer use. The loftiest expectation of those advocates was that the new technologies would be used to make practice more scientific and thus would serve to improve the knowledge base in social work.

GREAT EXPECTATIONS FOR RESEARCHERS

A few prescient observers foresaw that practitioners should not be expected to assume all the responsibility for adapting new knowledge and new technologies to practice. Realizing that between theory and practice lay a vast territory of development and preliminary testing, these innovators shifted the burden to researchers, requiring them to work more diligently in crafting and testing models of intervention before asking practitioners to consider using them. These attempts at social research and development focused on the missing engineering process between idea and application (chapter 5). Despite the convincing logic of using an R&D process and exporting already developed practice models to the field, the approach has been applied to the work of only a few productive groups of social work researchers.

The reasons are not obscure. The design and development process is lengthy, time-consuming, and expensive, often lasting years. It does not conform well to the demands of agencies and government for rapid answers to specific programmatic questions, nor does it entice young academic researchers, who are warned against long-term projects in their rush to gain tenure. Moreover, the funding for such work is increasingly difficult to secure, especially in the early phases of exploration and development. Even the later phases, involving dissemination and marketing, require years of sustained effort. And, given the individualistic nature of most social work research, engineering social work intervention can be a mammoth undertaking.

Quonset Huts of Lesser Knowledge

Social work schools make easy targets. This is especially true in the United States, where social work school-bashing has been a favorite sport for a wide range of participants over a long period of time. There are a number of characteristics of this institution that make it vulnerable to attack. Its origins are seen to be lowly . . . as is the social standing of its primary clientele (disproportionately drawn from the ranks of women and the working class), and it prepares students for one of the lesser professions. Its curriculum and academic standards are generally considered weak and its faculty and students less able than their counterparts elsewhere in the university. All of these elements make the social work school easy to pick on and difficult to defend.

Except for our replacement of the term "social work" for "education," the above is a direct quote from the opening paragraph of an insightful article about the predicament of schools of education, which has great relevance for a sister professional school, the graduate school of social work. Labaree (1998), an education professor at Michigan State University, offers an analysis of how the characteristics of educational knowledge—"lesser knowledge"—affect the work and social status of educational researchers in the university and in society, the way in which their enterprise is organized and functions. The parallels with social work are obvious.

SOFT KNOWLEDGE

First, compared to the natural sciences, educational knowledge is soft, not hard. The intellectual terrain is less clearly defined; the methodologies for verification are less clear; it is much more difficult to establish findings that are reproducible and can be defended against scientific challenge. Attention is focused on description and interpretation, and consequently, the producers of soft knowledge find themselves constantly rebuilding the foundations of their disciplines as reinterpretations of the fundamental issues are constantly offered. Second, researchers in soft knowledge, because the subject matter involves human beings, have embedded in their projects values and purposes that make their conclusions political and subject to reinterpretation by those with different views. Equally complicating is the fact that the institutions of education (read social welfare) are fundamentally political, reflecting social values. Despite efforts to make education "hard" by using the various technologies of research, like statistics, "the only causal claims educational research can make are constricted by a mass of qualifying clauses, which show that these claims are only valid within the artificial restrictions of a particular experimental setting or the complex peculiarities of a particular natural context" (5). For all these reasons, "there is little that researchers can do to construct towers of knowledge on the foundations of the work of others," as in the hard sciences. Instead, groups of researchers who share some values and interpretive approaches "construct Quonset huts of knowledge through a short-term effort of intellectual accumulation," but these are seen as structurally unsound by researchers who do not share their values (5).

APPLIED KNOWLEDGE

The hard sciences are oriented around the construction of theory, establishing abstract claims of a universal kind. This allows scientists to ignore the local context and focus on the general. It also allows them to take a lofty, long view of knowledge development and to have the luxury of not worrying about the immediate practicality of their work. The soft sciences, by contrast, focus on specific contexts and try to solve particular local problems. They are problem solving more than theory building. The applied disciplines, like education and social work, focus on a specific institutional public policy area and are called upon not just to elucidate but also to improve it. Researchers in applied fields are constrained to focus more on vexing current problems than on their own intellectual interests. There is less prestige in this local work.

EXCHANGE VALUE VERSUS USE VALUE

Education schools, like social work schools, Labaree argues, provide training in soft, applied knowledge that is low in exchange value but high in use value. By this he means that these professional schools offer training that is not of high enough prestige that it can be "exchanged" for a high-paying career, somewhat independent of the actual knowledge acquired. "Use value" refers to specific applied knowledge and skills that will enable students to get a job, independent of the prestige of the particular university. For example, in Labaree's terms, a bachelor's degree in comparative literature from Harvard has high exchange value (i.e., might get you into a good law school), but low use value (i.e., prepares you for no particular job), whereas an MSW from a major university has low exchange value but high use value. Researchers in soft knowledge in professional schools devoted to preparing students with high use-value educations rank low on the university's totem pole and in resource allocations.

ORGANIZATIONAL CONSEQUENCES

Because of these characteristics of their knowledge, the hard and soft disciplines evolved different structures for scientific work. In hard fields, because the knowledge pursued is cumulative and verifiable, researchers from many sites tend to be intellectually clustered on the same central scientific problems at the cutting edge of discovery. They share a common knowledge base and theoretical orientation; they have a tower of "givens" and accepted facts on which to build. Novices in hard fields must master the evolution of the knowledge base before they can work at the cutting edge. This produces a social structure that is hierarchical in two important ways: new knowledge is grafted onto old knowledge in a cumulative manner, and senior researchers are in positions of authority. Also, at the cutting edge of discovery, the research community has developed a shorthand way of communicating that assumes that all the active participants will know which issues have already been solved and understand the significance of the next scientific problems to be worked on. Nonexperts will not be able to understand or participate in this intellectual exchange.

Soft disciplines, by contrast, cannot so easily build on foundations of knowledge, because the foundations are not solid, always being reconstructed or reinterpreted. Furthermore, there is not one or a few central intellectual puzzles that are the focus of the research community's attention, but a broad spread of problems and issues not held together by any grand theoretical framework or hierarchy of knowledge or senior researchers. In fact, novice researchers can strike out on their own, paying little attention to the existing body of knowledge. The products of research, because they focus on local, immediate practical problems, are subject to review and commentary not just by other scholars but by the lay public.

> As a result, research work is spread thinly over a wide area as individuals and groups continually work at rethinking the most basic issues in the field and as they each pursue their own interpretive approaches. The resulting terrain is laid out in a series of rural dwellings and hamlets rather than in the kind of urban high rises erected by researchers in a field like physics. Novices in this setting find themselves inducted quickly because the field is wide open and no issues are considered closed off from reconsideration. Senior people have less control over the work of intellectual production because their own work is so easily subject to challenge. And the field is less turned in on itself because its boundaries are permeable, its body of knowledge non-esoteric, and its discourse diffused among a variety of divergent research communities.
>
> (Labaree 7)

The looseness of this lesser form of knowledge has its benefits: researchers are less subject to disciplinary, methodological, or hierarchical constraints; producing soft knowledge has enjoyed a renaissance as positivism has been attacked; and the applied nature of the inquiry attracts a general lay audience interested in the work. Thus, the very nature of the practical problems that social work researchers, like educational researchers, address impinges on their status in the university and society and affects their ability to produce a cumulative body of knowledge. The profession's knowledge base inevitably appears thin and fragmented and is the subject of endless political and ideological disputes. This presents graduate schools of social work with major challenges, as they should serve as guardians of the primary sites for developing knowledge for practice.

The Obligations of Graduate Schools

The social work research community is scattered across many academic "hamlets," each working on an array of substantive problem areas, not coordinated by any theoretical and intellectual structure. Many schools, in fact, have only a few active researchers, and they may be working alone on very different topics. The solutions to this lack of a coordinated research effort are not at all clear. But the graduate schools of social work will continue to carry the burden of protecting knowledge development in the profession.

The vitality of schools of social work is frequently confused with their expansion. Certainly, social work education has expanded in recent decades. There are now many more accredited MSW programs than 25 years ago, and the number of accredited undergraduate programs has mushroomed to more than 400. But this growth has taken place in academic institutions that are by design and mission devoted not to supporting a rich research environment but to preparing students to fill beginning social work roles. Even the increase in the number of doctoral programs has not necessarily led to a greatly enhanced research infrastructure because, according to a Task Force report (1991), the programs are not graduating more students and the quality of many of the programs is marginal. Doctoral education, which is the only site for the improvement of the profession's knowledge-building capacities, is often marginalized in schools of social work, which put most of their resources into MSW and BSW programs. Very few schools have doctoral education at the core of the educational enterprise, as it is in graduate programs in the hard sciences.

Without an organized and sustained research enterprise, social work will become a vulnerable profession unable to maintain credibility with universities, other professions, or society. It will be seen only as a derivative occupation—as Flexner suggested—providing services based on knowledge developed by others, rather than generating its own intellectual leadership in defining problems and designing solutions. If the profession fails to engage vigorously in the process of developing theories of practice and effective interventions, it will relinquish authority to do so to others outside. This abdication has already occurred in some areas where social work once had a major influence, such as corrections, juvenile justice, and income maintenance.

Professional knowledge does not emerge automatically from the helping activities of practitioners in forms that are verified or can be easily disseminated. Faculty who engage in practice research have the imposing task of understanding how skilled practitioners work, measuring their successes, and distilling from them testable intervention principles. The aptitudes, skills, and interests of practitioners and researchers are at times quite different, as we have noted in several previous chapters. Tensions between analytic and intuitive modes of reasoning and between academic skepticism and comforting faith in familiar practice exist. These differences are neither completely resolvable nor unhealthy for the profession, but social work cannot succeed by faith and ideology alone. The research community has an obligation to keep the profession liberated from its own provincialism and abreast of the latest research on effective practice.

Unlike some professions, social work has no Cold Springs Harbor Laboratory, Sloan-Kettering Institute, or Scripps Institution to conduct basic research. It benefits from no powerful cluster of private entrepreneurial design and development activity such as Silicon Valley. And social work has no counterpart to the National Institutes of Health. Unlike medicine, it has no research hospitals. Unlike law, it has no authoritative decision makers who preside over the interpretation and application of knowledge. If social work faculty are not expected to question, extend, and test the profession's knowledge, there is no one else who will. Thus, the obligations of knowledge development rest almost exclusively on the shoulders of faculty in a few graduate schools of social work.

Although teaching and scholarship must be joined in any university-affiliated professional school, rarely in social work are they evenly balanced. Part of the problem lies in the organizational culture and values of some schools where scholarship receives little support, beyond lip service. Part of that culture stems from the expansive intrusiveness of accreditation requirements that have little regard for the traditions, strengths, or obligations of the major research universities. Even among the BSW or MSW alumni of leading schools, few appreciate the importance and difficulty of knowledge development. Unfortunately, all too frequently, some faculty even harbor the belief that research actually detracts from the school's real work.

If social work wants to strengthen its intellectual leadership in universities and in the human services, the nature and role of research in the academic institution must be articulated better, understood better by the university, and more generously supported by the profession. Research and scholarship must play an enlarged role in social work schools if they and the profession are to survive and prosper. Failure to fulfill our obligations for knowledge development will undoubtedly threaten social work's long-term welfare and weaken its capacity to serve those in need. We trust that the lessons we can gain from past attempts to link science and practice will help prevent this from happening.

■ REFERENCES

Abbott, A. 1988. *The System of Professions*. Chicago: University of Chicago Press.
——. 1995. Boundaries of social work or social work of boundaries? *Social Service Review* 69:545–562.
——. 1999. *Department and Discipline: Chicago Sociology at One Hundred*. Chicago: University of Chicago Press.
Abels, P. 1972. Can computers do social work? *Social Work* 17 (5): 5–11.
Alexander, F. G. and Selesnick, S. T. 1966. *The History of Psychiatry*. New York: Harper and Row.
American Psychiatric Association. 1952. *Diagnostic and Statistical Manual of Mental Disorders*. Washington, DC: American Psychiatric Association.
——. 1968. *Diagnostic and Statistical Manual of Mental Disorders*, 2nd ed. Washington, DC: American Psychiatric Association.
——. 1980. *Diagnostic and Statistical Manual of Mental Disorders*, 3rd ed. Washington, DC: American Psychiatric Association.
——. 1987. *Diagnostic and Statistical Manual of Mental Disorders*, 3rd ed., rev. Washington, DC: American Psychiatric Association.
——. 1994. *Diagnostic and Statistical Manual of Mental Disorders*, 4th ed. Washington, DC: American Psychiatric Association.
——. 1997. Practice guidelines for the treatment of patients with Alzheimer's disease and other dementias of late life. Supplement of the *American Journal of Psychiatry* 154 (4): 1–63.
Anderson, C. M., Reiss, D. J., and Hogarty, G. E. 1986. *Schizophrenia and the Family*. New York: Guilford.
Andrews, D. A., Zinger, I., Hoge, R. D., Bonta, J., Gendreau, P., and Cullen, F. T. 1990.

Does correctional treatment work? A clinically relevant and psychologically informed meta-analysis. *Criminology* 28 (3): 369–386.

Aronson, J. L., Harre, R., and Way, E. C. 1995. *Realism Rescued: How Scientific Progress Is Possible*. Chicago: Open Court.

Aronson, S. H. and Sherwood, C. C. 1967. Researcher versus practitioner: Problems in social action research. *Social Work* 12:89–96.

Ashford, J. B. and LeCroy, C. W. 1991. Problem solving in social work practice: Implications for knowledge utilization. *Research on Social Work Practice* 1 (3): 306–318.

Asquith, S., Buist, M., and Loughran, N. 1998. *Children, Young People and Offending in Scotland: A Research Review*. Edinburgh: Scottish Office Central Research Unit.

Auslander, G. and Cohen, M. 1992. The role of computerized information systems in quality assurance in hospital social work departments. *Social Work in Health Care* 18:71–92.

Austin, D. 1997. The institutional development of social work education: The first 100 years—and beyond. *Journal of Social Work Education* 33 (3): 599–612.

——. 1999. A report on progress in the development of research resources in social work. *Research in Social Work Practice* 9:673–707.

Austin, L. N. 1948. Trends in differential treatment in social casework. *Journal of Social Casework* 23:203–211.

Bailey-Dempsey, C. and Reid, W. J. 1996. Intervention design and development: A case study. *Research on Social Work Practice* 6 (2): 208–228.

Baker, R. P. 1992. New technology in survey research: Computer-assisted personal interviewing (CAPI). *Social Science Computer Review* 10 (2): 145–157.

Baldi, J. J. 1971. Doctorates in social work, 1920–1968. *Journal of Education for Social Work* 7 (1): 11–22.

Barlett, H., Gordon, W., and Kadushin, A. 1964. Preface. In National Association of Social Workers, *Building Social Work Knowledge*. New York: National Association of Social Workers.

Barlow, D. H., ed. 1991. Diagnoses, dimensions, and *DSM–IV*: The science of classification [Special issue]. *Journal of Abnormal Psychology* 100 (3).

Barlow, D. H., Hayes, S. C., and Nelson, R. O. 1984. *The Scientist Practitioner: Research and Accountability in Clinical and Educational Settings*. New York: Pergamon.

Bateson, G., Jackson, D. D., Haley, J., and Weakland, J. H. 1956. Toward a theory of schizophrenia. *Behavior Science* 1:251–264.

Benbenishty, R. 1989. Designing computerized clinical information systems to monitor interventions on the agency level. *Computers in Human Services* 5 (1/2): 69–88.

——. 1996. Integrating research and practice: Time for a new agenda. *Research on Social Work Practice* 6 (1): 77–82.

——. 1997. Outcomes in the context of empirical practice. In E. J. Mullen and J.

Magnabosco, eds., *Outcome Measurement in the Human Services*, 198–208. Washington, DC: National Association of Social Workers.

Benbenishty, R. and Ben-Zaken, A. 1988. Computer-aided process of monitoring task-centered family interventions. *Social Work Research and Abstracts* 24:7–9.

Berlin, S. B., Mann, K. B., and Grossman, S. F. 1991. Task analysis of cognitive therapy for depression. *Social Work Research and Abstracts* 27:3–11.

Berlin, S. B. and Marsh, J. C. 1993. *Informing Practice Decisions*. New York: Macmillan.

Besa, D. 1994. Evaluating narrative family therapy using single-system research designs. *Research on Social Work Practice* 4 (3): 309–325.

Best and worst of the decade (review of the past 10 years) (Industry Trend or Event). 1998, September. *PC Computing* 11 (9): 58.

Beutler, L. E. 1991. Have all won and must all have prizes? Revisiting Luborsky et al.'s verdict. *Journal of Consulting and Clinical Psychology* 59 (2): 226–232.

Beyer, J. M. and Trice, H. M. 1982. The utilization process: A conceptual framework and synthesis of empirical findings. *Administrative Science Quarterly* 27:591–622.

Bhaskar, R. 1975. *A Realist Theory of Science*. Leeds, UK: Leeds Books Ltd.

Bloom, M. 1985. *Life Span Development*. New York: Macmillan.

Bloom, M., Fischer, J. and Orme, J. G. 1999. *Evaluating Practice*, 3rd ed. New York: Free Press.

Blythe, B. J. and Patterson, S. M. 1994. A review of intensive family preservation services research. *Social Work Research* 18:213–225.

Blythe, B. J. and Rodgers, A. Y. 1993. Evaluating our own practice: Past, present, and future trends. *Journal of Social Service Research* 18 (1/2): 101–119.

Blythe, B. J., Tripodi, T., and Briar, S. 1994. *Direct Practice Research in Human Services Agencies*. New York: Columbia University Press.

Bostwick, G. 1983. Starting out with microcomputers: Some practical advice for the novice. An interview. *Practice Digest* 6 (3): 6–10.

Boyd, L. H., Jr., Hylton, J. H., and Price, S. V. 1978. Computers in social work practice: A review. *Social Work* 23 (5): 368–371.

Boyd, L. H., Jr., Pruger, R., Chase, M. D., Clark, M., and Miller, L. S. 1981. A decision support system to increase equity. *Administration in Social Work* 5 (3–4): 83–96.

Brandt, L. 1906. Statistics of dependent families. *Proceedings of the National Conference of Charities and Corrections*, 440.

Breeding, W., Grishman, M., and Moreland, M. 1996. Implementation of computerized social work database assessments. *Social Work in Health Care* 23:81–98.

Bremner, R. H. 1956. Scientific philanthropy 1873–93. *Social Service Review* 30:21–34.

Brent, D. A. 1995. Risk factors for adolescent suicide and suicidal behavior: Mental and substance abuse disorders, family environmental factors, and life stress. *Suicide and Life-Threatening Behavior* 25:52–63.

Brent, E. 1988. New approaches to expert systems and artificial intelligence programming. *Social Science Computer Review* 6 (4): 569–578.

Briar, S. 1968a. The casework predicament. *Social Work* 13:5–11.

——. 1968b. The current crisis in social casework. In *Social Work Practice, 1967*, 19–33. Columbus, OH: National Conference on Social Welfare.

——. 1979. Incorporating research into education for clinical practice in social work: Toward a clinical science in social work. In A. Rubin and A. Rosenblatt, eds., *Sourcebook on Research Utilization*, 132–140. New York: Council on Social Work Education.

——. 1990. Empiricism in clinical practice: Present and future. In L. Videka-Sherman and W. J. Reid, eds., *Advances in Clinical Social Work Research*, 1–7. Washington, DC: National Association of Social Workers.

Briar, S. and Miller, H. 1971. *Problems and Issues in Social Casework*. New York: Columbia University Press.

Briar, S., Weissman, H., and Rubin, A., eds. 1981. *Research Utilization in Social Work Education*. New York: Council on Social Work Education.

Bronson, D. E. 1994. Is a scientist-practitioner model appropriate for direct social work practice? No. In W. W. Hudson and P. S. Nurius, eds., *Controversial Issues in Social Work Research*, 79–86. Boston: Allyn and Bacon.

Bronson, D. E. and Blythe, B. J. 1987. Computer support for single-case evaluation of practice. *Social Work Research and Abstracts* 23 (3): 10–13.

Brown, D. 1999, October 22. Medicine's growth curve: Healthy patients. *The Washington Post*, A1.

Broxmeyer, N. 1978. Practitioner-research in treating a borderline child. *Social Work Research and Abstracts* 14:5–lo.

Bunge, M. 1996. *Finding Philosophy in Social Science*. New Haven: Yale University Press.

Butterfield, W. H., ed. 1983. Computers for social work practitioners. [Special issue]. *Practice Digest* 6 (3).

——. 1987. Artificial intelligence: An introduction. *Computers in Human Services* 3 (1/2): 23–35.

——. 1995. Computer utilization. In R. L. Edwards, ed., *Encyclopedia of Social Work 1995*, 19th ed., 594–613. Washington, DC: National Association of Social Workers.

——. 1998. Human services and the information economy. *Computers in Human Services* 15:121–142.

Calhoun, K. S., Pilkonis, P. A., Moras, K., and Rehm, L. P. 1998. Empirically supported treatments: Implications for training. *Journal of Consulting and Clinical Psychology* 66:151–162.

Carlson, R. W. 1989. Capturing expertise in clinical information processing. *Computers in Human Services* 5 (1/2): 37–54.

Carnevale, D. 1999, November 12. Web services help professors detect plagiarism. *The Chronicle of Higher Education*, A49.

Caro, F. G. 1977. Research in social work: Program evaluation. In J. B. Turner, ed., *Encyclopedia of Social Work 1977*, 17th ed., 1199–1202. Washington, DC: National Association of Social Workers.

Caspi, J. 1995. *Task-Centered Model for Social Work Field Instruction*. Ph.D. diss., University of New York at Albany.

Caspi, J. and Reid, W. J. 1998. The task-centered model for field instruction: An innovative approach. *Journal of Social Work Education* 34:55–70.

Casselman, B. 1972. On the practitioner's orientation toward research. *Smith College Studies in Social Work* 42:211–233.

Chambless, D. L. and Hollon, S. D. 1998. Defining empirically supported theories. *Journal of Consulting and Clinical Psychology* 66 (1): 7–18.

Chandler, S. M. 1994. Is there an ethical responsibility to use practice methods with the best empirical evidence of effectiveness? No. In W. W. Hudson and P. S. Nurius, eds., *Controversial Issues in Social Work Research*, 105–111. Boston: Allyn and Bacon.

Clark, C. F. 1988. Computer applications in social work. *Social Work Research and Abstracts* 24 (1): 15–19.

Clark, L. A., ed. 1999. The concept of disorder: Evolutionary analysis and critique [Special issue]. *Journal of Abnormal Psychology* 108:371–472.

Cnaan, R. A. 1989. Introduction: Social work practice and information technology—an unestablished link. *Computers in Human Services* 5 (1/2): 1–15.

Cohen, J. 1976. A brief comment: Evaluating the effectiveness of an unspecified "casework" treatment in producing change. In J. Fischer, ed., *The Effectiveness of Social Casework*, 176–189. Springfield, IL: Charles C. Thomas.

Colcord, J. and Mann, Ruth Z. S. 1930. *The Long View: Papers and Addresses by Mary E. Richmond*. New York: Russell Sage Foundation.

Conboy, A., Auerbach, C., Beckerman, A., Schnall, D., and LaPorte, H. (2000). MSW student satisfaction with using single-system-design software to evaluate social work practice. *Research on Social Work Practice* 10:127–138.

Cook, T. D. 1987. Positivist critical multiplism. In W. R. Shadish and C. S. Relchardt, eds., *Evaluation Studies, 12*. Newbury Park, CA: Sage.

Corrigan, P. W., MacKain, S. J., and Liberman, R. P. 1994. Skill training modules—A strategy for dissemination and utilization of a rehabilitation innovation. In J. Rothman and E. J. Thomas, eds., *Intervention Research: Design and Development for Human Service*, 317–352. New York: Haworth.

Cowley, G. and Underwood, A. 1999, September 20. Finding the right Rx: Portable databases can make doctors more efficient. But this one helps them practice better medicine. *Newsweek*, 66–67.

Cronbach, L. 1975. Beyond the two disciplines of scientific psychology. *American Psychologist* 30:6–127.

Curtis, G. C. 1996. The scientific evaluation of new claims. *Research on Social Work Practice* 6:117–121.

Cwikel, J. G. and Cnaan, R. A. 1991. Ethical dilemmas in applying second-wave information technology to social work practice. *Social Work* 36 (2): 114–120.

Davis, H. 1976. Foreword. In E. Glaser, H. H. Abelson, and K. N. Garrison, eds., *Putting Knowledge to Use: A Distillation of the Literature Regarding Knowledge Transfer and Change.* Los Angeles: Human Interaction Research Institute.

Davis, I. P. and Reid, W. J. 1988. Event analysis in clinical practice and process research. *Social Casework* 69:298–288.

Dean, R. and Reinherz, H. 1986. Psychodynamic practice and single system design: The odd couple. *Journal of Social Work Education* 22:71–81.

Department of Health and Human Services. 2000. http://www.acf.dhhs.gov/programs/oss/sacwis/!sacwis.htm

DeRubeis, R. J. and Crits-Christoph, P. 1998. Empirically supported individual and group psychological treatments for adult mental disorders. *Journal of Consulting and Clinical Psychology* 66 (1): 37–52.

DeSchmidt, A. and Gorey, K. M. 1997. Unpublished social work research: Systematic replication of a recent meta-analysis of published intervention effectiveness research. *Social Work Research* 21 (1): 58–62.

Dewees, M. 1999. The application of social constructionist principles to teaching in social work practice in mental health. *Journal of Teaching in Social Work* 19 (1/2): 31–46.

Dewey, J. 1938. *Logic: The Theory of Inquiry.* New York: Holt, Rinehart and Winston.

Donahue, K. M. 1996. *Developing a Task-Centered Mediation Model.* Ph.D. diss., State University of New York at Albany.

Dore, M. M. 1999. The retail method of social work: The role of the New York school in the development of clinical practice. *Social Service Review* 73 (2): 168–190.

Dvorak, J. C. 1998, December 1. What ever happened to . . . the first personal computer? *Computer Shopper*, 422.

Eaton, J. 1962. Symbolic and substantive evaluative research. *Administrative Science Quarterly* 6:421–442.

Edwards, R. L., ed. 1995. *Encyclopedia of Social Work*, 19th ed. Washington, DC: National Association of Social Workers.

Elliott, R. 1984. A discovery-oriented approach to significant change in psychotherapy: Interpersonal process recall and comprehensive process analysis. In L. N. Rice and L. S. Greenberg, eds., *Patterns of Change: Intensive Analysis of Psychotherapy Process*, 249–286. New York: Guilford.

Emlen, A., et al. 1978. *Overcoming Barriers to Planning for Children in Foster Care.* Portland, OR: Regional Research Institute for Human Services, Portland State University.

Emmelkamp, P. M. G., deHaan, E., and Hoodguin, C. A. L. 1990. Marital adjustments and obsessive-compulsive disorder. *British Journal of Psychiatry* 156:55–60.

The Encyclopedia Americana. 1953. 30 vols. New York: Americana Corp.

Eriksson, I. 1990. Use of research and different discourses in social work. *Knowledge: Creation, Diffusion, Utilization* 12 (1): 14–26.

Evans, G. 1889. Scientific charity. *Proceedings of the National Conference of Charities and Correction*, 24.

Fairweather, G. W. 1967. *Methods for Experimental Social Innovation*. New York: Wiley.

Family Service Association of America. 1953. Scope and methods of the family service agency: Report of the committee on methods and scope. New York: Family Service Association of America.

Fanshel, D. 1977. Parental visiting of children in foster care: A computerized study. *Social Work Research and Abstracts* 13:2–10.

——, ed. 1980. *Future of Social Work Research*. Washington, DC: National Association of Social Workers.

Faul, A. C., McMurtry, S. L., and Hudson, W. W. 2001. Can empirical clinical practice techniques improve social work outcomes? *Research on Social Work Practice* 11:277–299.

Fawcett, S. B., Suarez-Balcazar, Y., Balcazar, F. E., White, G. W., Paine, A. L., Blanchard, K. A., and Embree, M. G. 1994. Conducting intervention research: The design and development process. In J. Rothman and E. J. Thomas, eds., *Intervention Research: Design and Development for Human Service*, 25–54. New York: Haworth.

Finn, J. 1988. Microcomputers in private, nonprofit agencies: A survey of trends and training requirements. *Social Work Research and Abstracts* 24 (1): 10–14.

Finn, J. and Lavitt, M. 1994. Computer-based self-help groups for sexual abuse survivors. *Social Work with Groups* 17 (1/2): 21–46.

Finnegan, D. J., Ivanoff, A., and Smyth, N. J. 1991. The computer application explosion: What practitioners and clinical managers need to know. *Computers in Human Services* 8 (2): 1–19.

Fischer, J. 1973. Is casework effective?: A review. *Social Work* 18:5–20.

——, ed. 1976. *The Effectiveness of Social Casework*. Springfield, IL: Charles Thomas.

——. 1978. *Effective Casework Practice: An Eclectic Approach*. New York: McGraw-Hill.

——. 1981. The revolution in social work. *Social Work* 26:199–207.

——. 1992. Empirically based practice: The end of ideology? *Journal of Social Service Research* 18 (1/2): 19–64.

Fischer, J. and Corcoran, K. 1994. *Measures for Clinical Practice*, 2nd ed. Vol. 1: Couples, Families, and Children. New York: Free Press.

Fishman, D. B. 1988. Pragmatic behaviorism. Saving and nurturing the baby. In D. B. Fishman, E. Rotgers, and C. M. Franks, eds., *Paradigms in Behavior Therapy: Present and Promise*. New York: Springer.

Fishman, D. B., Rotgers, E., and Franks, C. M., eds. 1988. *Paradigms in Behavior Therapy: Present and Promise*. New York: Springer.

Flexner, A. 1915. Is social work a profession? *Proceedings of the National Conference of Charities and Correction*. Chicago: Hildmann, 577–590.

Fortune, A. E. 1981. Communication processes in social work practice. *Social Service Review* 55:93–128.

Fortune, A. E. and Reid, W. J. 1999. *Research in Social Work*, 3rd ed. New York: Columbia University Press.

Frankfurter, F. 1915. Social work and professional training. *Proceedings of the National Conference of Charities and Correction*. Chicago: Hildmann, 591–596.

Franklin, C., Nowicki, J., Trapp, J., Schwab, A., and Peterson, J. 1993. A computerized assessment system for brief, crisis-oriented youth services. *Families in Society* 74:602–616.

Fraser, M. W., Lewis, R. E., and Norman, J. L. 1991. Research education in M.S.W. programs: An exploratory analysis. *Journal of Teaching in Social Work* 4 (2): 83–103.

Fraser, M. W. and Nelson, K. E. 1997. Effectiveness of family preservation services. *Social Work Research* 21:138–154.

Fraser, M., Taylor, M. J., Jackson, R., and O'Jack, J. 1991. Social work and science: Many ways of knowing? *Social Work Research and Abstracts* 27 (4): 5–15.

Freedman, J. and Combs, G. 1996. *The Social Construction of Preferred Realities*. New York: Norton.

Friedman, R. M. 1980. The use of computers in the treatment of children. *Child Welfare* 59 (3): 152–159.

Fuller, T. K. 1970. Computer utility in social work. *Social Casework* 51 (10): 606–611.

Gambrill, E. 2000. Evidence-based practice: Implications for knowledge development and use in social work. Paper presented at Toward the Development of Evidence-Based Practice for Social Work Intervention: A Working Conference, May 3–5, 2000. George Warren Brown School of Social Work, Washington University, St. Louis, MO.

Garvey, W. D. and Griffith, B. C. 1967. Communication in a science: The system and its modification. In A. deReuck and J. Knight, eds. *Communication in Science*, 6–36. Boston: Little, Brown.

Geismar, L. L. 1982. Comments on the "Obsolete scientific imperative in social work research." *Social Service Review* 56:311–312.

Geiss, G. 1983. Some thoughts about the future: Information technology and social work practice. *Practice Digest* 6 (3): 33–35.

Gelman, S. R., Pollack, D., and Weiner, A. 1999. Confidentiality of social work records in the computer age. *Social Work* 44 (3): 243–252.

Gergen, K. J. 1985. The social constructionist movement in modern psychology. *American Psychologist* 40:260–275.

Germain, C. 1970. Casework and science: A historical encounter. In R. W. Roberts and R. H. Nee, eds. *Theories of Social Casework*, 3–32. Chicago: University of Chicago Press.

Gibbs, L. 1990. Using online databases to guide practice and research. *Computers in Human Services* 6 (1/2/3): 97–116.

Gingerich, W. J. 1984. Generalizing single-case evaluation from classroom to practice setting. *Journal of Education for Social Work* 20:74–82.

——. 1990a. Expert systems and their potential uses in social work. *Families in Society: The Journal of Contemporary Human Services* 71 (4): 220–228.

——. 1990b. Rethinking single-case evaluation. In L. Videka-Sherman and W. J. Reid, eds. *Advances in Clinical Social Work Research*, 11–24. Washington, DC: National Association of Social Workers.

——. 1995. Expert systems. In R. L. Edwards, ed., *Encyclopedia of Social Work 1995*, 19th ed., 917–925. Washington, DC: National Association of Social Workers.

Glaser, B. G. and Strauss, A. L. 1967. *The Discovery of Grounded Theory: Strategies for Qualitative Research*. Chicago: Aldine.

Glaser, E., Abelson, H. H., and Garrison, K. N., eds. 1976. *Putting Knowledge to Use: A Distillation of the Literature Regarding Knowledge Transfer and Change*. Los Angeles: Human Interaction Research Institute.

——. 1983. *Putting Knowledge to Use: Facilitating the Diffusion of Knowledge and the Implementation of Planned Change*. San Francisco: Jossey-Bass.

Glaser, E. M. and Taylor, S. H. 1973. Factors influencing the success of applied research. *American Psychologist* 28 (2): 140–146.

Goldstein, E. G. 1998. Psychology and object relations theory. In R. Dorfman, ed., *Paradigms of Clinical Social Work*, vol. 2. New York: Brunner/Mazel.

Goode, W. 1969. The theoretical limits of professionalism. In A. Etzioni, ed., *The Semi-Professions and Their Organization*. New York: Free Press.

Goodman, H., Gingerich, W. J., and de Shazer, S. 1989. Briefer: An expert system for clinical practice. *Computers in Human Services* 5 (1/2): 53–68.

Gorey, K. M. 1996. Effectiveness of social work intervention research: Internal versus external evaluations. *Social Work Research* 20 (2): 119–128.

Gorey, K. M. and Thyer, B. A. 1998. Differential effectiveness of prevalent social work practice models: A meta-analysis. *Social Work* 43 (3): 269–279.

Gottman, J. M. and Markman, H. J. 1978. Experimental designs in psychotherapy research. In S. L. Garfield and A. E. Bergin, eds., *Handbook of Psychotherapy and Behavior Change*, 2nd ed. New York: Wiley.

Graebner, W. 1987. *The Engineering of Consent: Democracy and Authority in Twentieth-Century America*. Madison: University of Wisconsin Press.

Grant, G. B. and Grobman, L. M., eds. 1998. *The Social Worker's Internet Handbook*. Harrisburg, PA: White Hat Communications.

Grasso, A. J. and Epstein, I. 1987. Management by measurement: Organizational dilemmas and opportunities. *Administration in Social Work* 11 (3/4): 89–100.

——. 1989. The Boysville experience: Integrating practice decision-making, program evaluation, and management information. *Computers in Human Services* 4:84–95.

——, eds. 1992. *Research Utilization in the Social Services: Innovations for Practice and Administration*. New York: Haworth.

Greenberg, L. S. 1984. Task analysis: The general approach. In L. N. Rice and L. S. Greenberg, eds., *Patterns of Change: Intensive Analysis of Psychotherapy*, 67–123. New York: Guilford.

Greene, G. J. and Jensen, C. 1996. A constructive perspective on clinical social work practice with ethnically diverse clients. *Social Work* 41 (2): 172–181.

Greenwood, E. 1957. Attributes of a profession. *Social Work* 2 (3): 45–55. Reprinted in Gilbert, N. and Specht, H., eds. 1976. *The Emergence of Social Welfare and Social Work*, 302–318. Itasca, IL: Peacock.

Grob, G. N. 1991. Origins of *DSM–I*: A study in appearance and reality. *American Journal of Psychiatry* 148:421–431.

Gross, P. R. and Levitt, N. 1994. *Higher Superstition*. Baltimore: The John Hopkins University Press.

Guba, E. G. 1968. Development, diffusion and evaluation. In T. L. Eiddell and J. M. Kitchel, eds., *Knowledge Production and Utilization*, 37–63. University Council for Educational Administration (Columbus, OH) and Center for Advanced Study of Educational Administration (University of Oregon).

——. 1990. The alternative paradigm dialog. In E. G. Guba, ed., *The Paradigm Dialog*, 17–27. Thousand Oaks, CA: Sage.

Guba, E. G. and Lincoln, Y. 1982. *Effective Evaluation*. San Francisco: Jossey-Bass.

Haack, S. 1996. Concern for truth: What it means, why it matters. In P. R. Gross, N. Levitt, and M. W. Lewis, eds., *The Flight from Science and Reason*, 57–63. New York: The New York Academy of Sciences.

Hacking, I. 1999. *The Social Construction of What?* Cambridge: Harvard University Press.

Hall, F. S., ed. 1965. *Encyclopedia of Social Work*, 1st ed. New York: Russell Sage Foundation.

Hamilton, G. 1951. *Theory and Practice of Social Case Work*, 2nd ed. New York: Columbia University Press.

Harrison, W. D. 1994. The inevitability of integrated methods. In E. Sherman and W. J. Reid, eds., *Qualitative Research in Social Work*, 409–422. New York: Columbia University Press.

Havelock, R. G. 1968. Dissemination and translation roles. In T. L. Eiddell and J. M. Kitchel, eds., *Knowledge Production and Utilization*, 64–119. University Council for Educational Administration (Columbus, OH) and Center for Advanced Study of Educational Administration (University of Oregon).

——. 1969. *Planning for Innovation Through Dissemination and Utilization of Knowledge*. Ann Arbor: Center for Research on Utilization of Scientific Knowledge, University of Michigan.

Haworth, G. O. 1984. Social work research, practice, and paradigms. *Social Service Review* 58:343–357.

Hayes, S. C., Nelson, R. D., and Jarrett, R. B. 1987. The treatment utility of assessment. *American Psychologist* 42:63–74.

Heineman, M. P. 1981. The obsolete imperative in social work research. *Social Service Review* 55:371–397.

——. 1994. Science, not scientism: The robustness of naturalistic clinical research. In E. Sherman and W. J. Reid, eds., *Qualitative Research in Social Work*, 71–88. New York: Columbia University Press.

Heineman-Pieper, M. 1989. The heuristic paradigm: A unifying and comprehensive approach to social work research. *Smith College Studies* 60:8–34.

Hepworth, D., Rooney, R., and Larson, N. 1997. *Direct Social Work Practice: Theory and Skills*, 5th ed. New York: Brooks/Cole.

Hess, P. and Mullen, E. J., eds. 1995. *Practitioner-Researcher Partnerships: Building Knowledge from, in, and for Practice*. Washington, DC: National Association of Social Workers.

Hesse, M. 1980. *Revolutions and Reconstructions in the Philosophy of Science*. Bloomington: Indiana University Press.

Hill, A. L. and Scanlon, C. R. 1998. Opening space and the two-story technique. *Journal of Family Psychotherapy* 9 (1): 75–79.

Hogarty, G. E. 1993. Prevention of relapse in chronic schizophrenic patients. *Clinical Psychiatry* 54:3.

Holden, G., Bearlson, D., Rode, D., Rosenberg, G., and Fishman, M. 1999. Evaluating the effects of a virtual environment (STARBRIGHT World) with hospitalized children. *Research on Social Work Practice* 9:365–382.

Holden, G., Rosenberg, G., and Meenaghan, T. In press. Information for social work practice: Observations regarding the role of the World Wide Web. *Social Work in Health Care*.

Hollis, F. 1963. Contemporary issues for caseworkers. In H. J. Parad and R. R. Miller, eds., *Ego-Oriented Casework*, 83–98. New York: Family Service Association of America.

——. 1964. *Casework: A Psychosocial Therapy*. New York: Random House.

Hoshino, G. and McDonald, T. P. 1975. Agencies in the computer age. *Social Work* 20 (1): 10–14.

Howard, M. O. and Jenson, J. M. 1999. Barriers to development, utilization, and evaluation of social work practice guidelines: Toward an action plan for social work. *Research on Social Work Practice* 9 (3): 347–364.

——. In press. Clinical guidelines and evidence-based practice in medicine, psychology, and allied professions. In E. Proctor and A. Rosen, eds., *Developing Practice Guidelines for Social Work Intervention*. New York: Columbia University Press.

Hudson, W. W. 1976. Special problems in the assessment of growth and deterioration. In J. Fischer, ed., *The Effectiveness of Social Casework*. Springfield, IL: Charles C. Thomas.

———. 1982a. *The Clinical Measurement Package: A Field Manual.* Homewood, IL: Dorsey.

———. 1982b. Scientific imperatives in social work research and practice. *Social Service Review* 56:246–58.

———. 1990. Computer-based clinical practice: Present status and future possibilities. In L. Videka-Sherman and W. J. Reid, eds., *Advances in Clinical Social Work Research.* Washington, DC: National Association of Social Workers.

Icard, L. D., Schilling, R. F., and El-Bassel, N. 1995. Reducing HIV infection among African Americans by targeting the African American family. *Social Work Research* 19:153–163.

Ivanoff, A., Blythe, B. J., and Briar, S. 1997. What's the story, morning glory? *Social Work Research* 21 (3): 194–196.

Ivanoff, A., Blythe, B. and Tripodi, T. 1994. *Involuntary Clients in Social Work Practice: A Research-Based Approach.* Hawthorne, NY: Aldine de Gruyter.

Ivanoff, A. and Riedel, M. 1995. Suicide. In R. L. Edwards, ed., *Encyclopedia of Social Work 1995*, 19th ed., 2358–2372. Washington, DC: National Association of Social Workers.

Izzo, R. L. and Ross, R. R. 1990. Meta-analysis of rehabilitation programs for juvenile delinquents: A brief report. *Criminal Justice and Behavior* 17 (1): 134–142.

Jaffe, E. D. 1979. Computers in child placement planning. *Social Work* 24 (5): 380–385.

Jeter, H. R. 1937. Research in social work. In R. Kurtz, ed., *Social Work Year Book 1937*, 4th ed., 419. New York: Russell Sage Foundation.

Kadushin, A. 1959. The knowledge base of social work. In A. J. Kahn, ed., *Issues in American Social Work*, 39–79. New York: Columbia University Press.

Kagle, J. 1982. Using single-subject measures in practice decisions: Systematic documentation or distortion? *Arete* 7 (2): 1–9.

Kammerman, S. B., Dolgoff, R., Getzel, G., and Nelsen, J. 1973. Knowledge for practice: Social science in social work. In A. J. Kahn, ed., *Shaping the New Social Work*, 97–146. New York: Columbia University Press.

Karls, J. and Wandrei, K. 1988. Person-in-Environment (PIE): A system for desribing, classifying, and coding problems of social functioning. Unpublished paper.

———. 1994. *Person-in-Environment System: The PIE Classification System for Social Functioning Problems.* Washington, DC: National Association of Social Workers.

Karpf, M. J. 1931. *The Scientific Basis of Social Work.* New York: Columbia University Press.

Kaye, L. 1991. A social work administration model curriculum in computer technolgy and information management. *Journal of Teaching in Social Work* 5:49–63.

Kazdin, A. E. and Bass, D. 1989. Power to detect differences between alternative treatments in comparative psychotherapy outcome research. *Journal of Consulting and Clinical Psychology* 57:138–147.

Kazi, M. A. F., Mantysaari, M., and Rostila, I. 1997. Promoting the use of single-case designs: Social work experiences from England and Finland. *Research on Social Work Practice* 7 (3): 311–328.

Kazi, M. A. F. and Wilson, J. 1996a. Applying single-case evaluation methodology in a British social work agency. *Research on Social Work Practice* 6 (1): 5–26.

——. 1996b. Applying single-case evaluation in social work. *British Journal of Social Work* 26:699–717.

Kellogg, P. U. 1914. *Pittsburgh District: Civic Frontage*. New York: Russell Sage Foundation.

Ketefian, S. 1975. Application of selected nursing research findings in nursing practice: A pilot study. *Nursing Research* 24 (2): 91.

Kiernan, V. 1999a, July 23. Editor of 'Science' voices doubts on NIH's proposed on-line archive. *The Chronicle of Higher Education*, A41.

——. 1999b, September 10. NIH Proceeds with on-line archive for papers in the life sciences. *The Chronicle of Higher Education*, A33.

Kilgore, D. K. 1995. *Task-Centered Group Treatment of Sex Offenders: A Developmental Study*. Ph.D. diss., State University of New York at Albany.

Kirk, S. A. 1979. Understanding research utilization in social work. In A. Rubin and A. Rosenblatt, eds., *Sourcebook on Research Utilization*, 3–15. New York: Council on Social Work Education.

——. 1990. Research utilization: The substructure of belief. In L. Videka-Sherman and W. J. Reid, eds., *Advances in Clinical Social Work Research*, 233–250. Washington, DC: National Association of Social Workers.

——. 1999. Good intentions are not enough: Practice guidelines for social work. *Research on Social Work Practice* 9:302–310.

Kirk, S. A. and Fischer, J. 1976. Do social workers understand research? *Journal of Education for Social Work* 12:63–70.

Kirk, S. A. and Kutchins, H. 1988. Deliberate misdiagnosis in mental health practice. *Social Service Review* 62 (2): 225–237.

——. 1992. *The Selling of DSM: The Rhetoric of Science in Psychiatry*. Hawthorne, NY: Aldine de Gruyter.

Kirk, S. A., Osmalov, M., and Fischer, J. 1976. Social workers' involvement in research. *Social Work* 21:121–124.

Kirk, S. A. and Penka, C. E. 1992. Research utilization and MSW education: A decade of progress? In A. J. Grasso and I. Epstein, eds., *Research Utilization in the Social Services*, 407–419. New York: Haworth.

Kirk, S. A., Siporin, M., and Kutchins, H. 1989. The prognosis for social work diagnosis. *Social Casework* 70 (5): 295–304.

Kirk, S. A., Wakefield, J., Hsieh, D., and Pottick, K. 1999. Social context and social workers' judgments of mental disorder. *Social Service Review* 73:82–104.

Kitchner, P. 1993. *The Advancement of Science*. New York: Oxford University Press.

Klein, P. 1931. Mary Richmond's formulation of a new science. In S. Rice, ed., *Methods in Social Science*. Chicago: University of Chicago Press.

Klein, W. C. and Bloom, M. 1995. Practice wisdom. *Social Work* 40 (2): 799–807.

Kutchins, H. and Kirk, S. A. 1988. The business of diagnosis: *DSM–III* and clinical social work. *Social Work* 33 (3): 215–220.

———. 1997. *Making Us Crazy. DSM: The Psychiatric Bible and the Creation of Mental Disorders.* New York: Free Press.

Labaree, D. 1998. Education researchers: Living with a lesser form of knowledge. *Educational Researcher* 27:4–12.

Lakatos, I. 1972. Falsification and the methodology of scientific research programs. In I. Lakatos and A. Musgrave, eds., *Criticisms and the Growth of Knowledge.* Cambridge: Cambridge University Press.

Lambert, M. J. and Hill, C. E. 1994. Assessing psychotherapy outcomes and processes. In A. E. Bergin and S. L. Garfield, eds., *Handbook of Psychotherapy and Behavior Change,* 72–113. New York: Wiley.

LeCroy, C. W., ed. 1994. *Handbook of Child and Adolescent Treatment Manuals.* New York: Lexington.

Lee, P. 1915. Committee Report: The professional basis of social work. *Proceedings of the National Conference of Charities and Correction,* 596–606. Chicago: Hildmann.

Lehman, A. F., Steinwachs, D. M., and the co-investigators of the PORT project. 1998. At issue: Translating research into practice: The schizophrenia patient outcomes research team (PORT) treatment recommendations. *Schizophrenia Bulletin* 24:1–9.

Leviton, L. C. and Boruch, R. F. 1980. Illustrative case studies. In R. F. Boruch and D. S. Cordray, eds., *An Appraisal of Educational Program Evaluations: Federal, State, and Local Levels.* Washington, DC: U.S. Department of Education.

Leviton, L. C. and Hughes, E. F. 1981. Research on the utilization of evaluations: A review and synthesis. *Evaluation Review* 5:525–548.

Light, R. and Pillemer, D. B. 1984. *Summing up: The Science of Reviewing Research.* Cambridge: Harvard University Press.

Lincoln, Y. and Guba, E. 1985. *Naturalistic Inquiry.* Beverly Hills, CA: Sage.

Lipsey, L. W. 1992. Juvenile delinquency treatment: A meta-analytic inquiry into the variability of effects. In T. D. Cook, H. Cooper, D. S. Conday, H. Hartmann, L. V. Hedges, T. A. Louis, and F. Mosteller, eds., *Meta-Analysis for Explanation: A Casebook,* 83–127. New York: Russell Sage Foundation.

Loch, C. S. 1899. Christian charity and political economy. *Charity Organization Review* 6:10–20.

Lowell, J. S. 1884. *Public Relief and Public Charity.* New York: Knickerbocker Press.

Luborsky, L. 1995. Are common factors across different psychotherapies the main explanation for the dodo bird verdict that "Everyone has won so all shall have prizes?" *Clinical Psychological Science and Practice* 2:106–109.

Luborsky, L., Singer, B., and Luborsky, L. 1975. Comparative studies of psychotherapy. *Archives of General Psychiatry* 32:995–1008.

Lubove, R. 1965. *The Professional Altruist*. Cambridge: Harvard University Press.

Luhrmann, T. M. 2000. *Of Two Minds: The Growing Disorder in American Psychiatry*. New York: Knopf.

Maas, H. S. 1977. Research in social work. In J. B. Turner, ed., *Encyclopedia of Social Work 1977*, 17th ed., 1183–1194. Washington, DC: National Association of Social Workers.

MacDonald, G., Sheldon, B. and Gillespie, J. 1992. Contemporary studies of the effectiveness of social work. *British Journal of Social Work* 22 (6): 625–643.

MacDonald, M. E. 1960. Social work research: A perspective. In N. A. Polanski, ed., *Social Work Research*, 1–23. Chicago: University of Chicago Press.

MacDonald, M. E. 1966. Reunion at Vocational High. *Social Service Review* 40:175–189.

MacIver, R. M. 1931. *The Contribution of Sociology to Social Work*. New York: D. Appleton-Century.

Marino, R., Green, G. R., and Young, E. 1998. Beyond the scientist-practitioner model's failure to thrive: Social workers' participation in agency-based research activities. *Social Work Research* 22:188–191.

Mattaini, M. 1993. *More Than a Thousand Words: Graphics for Clinical Practice*. Washington, DC: National Association of Social Workers.

Mattaini, M. and Kirk, S. A. 1991. Assessing assessment in social work. *Social Work* 36 (3): 260–266.

McQuaide, S. 1999. A social worker's use of the *Diagnostic and Statistical Manual*. *Families in Society* 80:410–416.

Meier, A. 1997. Inventing new models of social support groups: A feasibility study of an online stress management support group for social workers. *Social Work with Groups* 20 (4): 35–53.

Meyer, C. H. 1972. Practice on the microsystem level. In E. J. Mullen and J. Dumpson, eds., *Evaluation of Social Intervention*, 158–190. London: Jossey-Bass.

——. 1973. Direct services in new and old contexts. In A. J. Kahn, ed., *Shaping the New Social Work*, 26–54. New York: Columbia University Press.

Meyer, H., Borgatta, E., and Jones, W. 1965. *Girls at Vocational High*. New York: Russell Sage Foundation.

Meyers, T., Miltenberger, R. G., and Suda, K. 1998. A survey of the use of facilitated communication in community agencies serving persons with developmental disabilities. *Behavioral Interventions* 13:135–146.

Miller, M. J. 1981. Original title: Introduction of the PC: 1981. (Looking Back). *PC Magazine* 16 (6) (March 25, 1997): 108–111.

Mills, L. 1998. *The Heart of Intimate Abuse*. New York: Springer.

——. 1999. Killing her softly: Intimate abuse and the violence of state intervention. *Harvard Law Review* 113:550–613.

Millstein, K. H., Regan, J., and Reinherz, H. 1990, March. Can training in single subject design generalize to non-behavioral practice? Paper presented at the

Annual Program Meeting of the Council of Social Work Education, Reno, NV.

Minahan, A., ed. 1987. *Encyclopedia of Social Work*, 18th ed. Washington, DC: National Association of Social Workers.

Mirowsky, J. and Ross, C. 1989a. Psychiatric diagnosis as reified measurement. *Journal of Health and Social Behavior* 30:11–25.

———. 1989b. *Social Causes of Psychological Distress*. Hawthorne, NY: Aldine de Gruyter.

Monnickendam, M. 1999. Computer systems that work: A review of variables associated with system use. *Journal of Social Service Research* 26:71–94.

Morrissey, J. P. and Jones, W. C. 1977. Research in social work: Interorganizational analysis. In J. B. Turner, ed., *Encyclopedia of Social Work 1977*, 17th ed., 1194–1199. Washington, DC: National Association of Social Workers.

Morrow-Bradley, C. and Elliott, R. 1986. Utilization of psychotherapy research by practicing psychotherapists. *American Psychologist* 41:188–197.

Mullen, E. J. 1978. The construction of personal models for effective practice: A method for utilizing research findings to guide social interventions. *Journal of Social Service Research* 2 (1): 45–63.

———. 1983. Personal practice models. In A. Rosenblatt and D. Waldfogel, eds., *Handbook of Clinical Social Work*, 623–649. San Francisco: Jossey-Bass.

———. 1994. Design of social intervention. In J. Rothman and E. J. Thomas, eds., *Intervention Research*, 163–193. New York: Haworth.

Mullen, E. J. and Bacon, E. In press. Practitioner adoption and implementation of evidence-based effective treatment and issues of quality control. In E. Proctor and A. Rosen, eds., *Developing Practice Guidelines for Social Work Intervention*. New York: Columbia University Press.

Mullen, E. J., Chazin, R., and Feldstein, D. 1972. Services for the newly dependent: An assessment. *Social Service Review* 46:309–322.

Mullen, E. J., Dumpson, J. R., et al. 1972. *Evaluation of Social Intervention*. San Francisco: Jossey-Bass.

Mullen, E. J. and Schuerman, J. 1990. Expert systems and the development of knowledge in social welfare. In L. Videka-Sherman and W. J. Reid, eds., *Advances in Clinical Social Work Research*, 67–83. Washington, DC: National Association of Social Workers.

Munson, F. C. and Pelz, D. C. 1982. *Innovating in Organizations: A Conceptual Framework*. Ann Arbor: University of Michigan, Institute for Social Research.

Murphy, J. W. and Pardeck, J. T. 1992. Computerization and the dehumanization of social services. *Administration in Social Work* 16 (2): 61–72.

Mutschler, E. 1984. Evaluating practice: A study of research utilization by practitioners. *Social Work* 29:332–337.

Mutschler, E. and Jayaratne, S. 1993. Integration of information technology and single-system designs: Issues and promises. In M. Bloom, ed., *Single-System Designs in the Social Services: Issues and Options for the 1990s*, 121–145. New York: Haworth.

Myers, L. L. and Thyer, B. A. 1997. Should social work clients have the right to effective treatment? *Social Work* 42 (3): 288–299.

Myrick, H. L. 1928. *Interviews: A Study of the Methods of Analyzing and Recording Social Case Work Interviews*. New York: Herald Nathan Press.

Nagel, T. 1997. *The Last Word*. New York: Oxford University Press.

Naleppa, M. J. 1995. *Task-Centered Case Management with Elderly: Developing a Practice Model*. Ph.D. diss., State University of New York at Albany.

Naleppa, M. J. and Reid, W. J. 1998. Task-centered case management for the elderly: Developing a practice model. *Research on Social Work Practice* 8:63–85.

National Association of Social Workers. 1964. *Building Social Work Knowledge: Report of a Conference*. New York: National Association of Social Workers.

——. 2001. Caution urged before web counseling. *NASW News* (January):5.

Nelsen, J. C. 1978. Use of communication theory in single-subject research. *Social Work Research and Abstracts* 14 (12): 12–19.

Noble, J. H., Jr. 1971. Protecting the public's privacy in computerized health and welfare systems. *Social Work* 16 (1): 35–41.

Nurius, P. S. and Hudson, W. W. 1988a. Workers, clients, and computers. *Computers in Human Services* 4 (1/2): 71–83.

——. 1988b. Computer-based practice: Future dream or current technology? *Social Work* 33 (4): 357–362.

——. 1993. *Human Services: Practice, Evaluation, and Computers*. Pacific Grove, CA: Brooks/Cole.

Nurius, P. S., Hooyman, N., and Nicoll, A. E. 1991. Computers in agencies: A survey baseline and planning implications. *Journal of Social Service Research* 14 (3/4): 141–155.

Olson, D. H. 1972. Empirically unbinding the double bind: Review of research and conceptual reformulations. *Family Process* II:69–94.

O'Neill, J. V. 2001. Online therapy on verge of major launch. *NASW News* (January):5.

Orlinsky, D. E. and Howard, K. I. 1986. Process outcome in psychotherapy. In S. L. Garfield and A. E. Bergin, eds., *Handbook of Psychotherapy and Behavior Change*. New York: Wiley.

Paine, S. C., Bellamy, B. T., and Wilcox, B. 1984. *Human Services That Work: From Innovation to Standard Practice*. Beverly Hills, CA: Sage.

Pardeck, J. T. 1997. Computer technology in clinical practice: A critical analysis. *Social Work and Social Sciences Review* 7:101–111.

——. 1998. Rationalizing decision making through computer technology: A critical appraisal. *Journal of Health and Social Policy* 9:19–29.

Pardeck, J. T. and Schulte, R. S. 1990. Computers in social intervention: Implications for professional social work practice and education. *Family Therapy* 17 (2): 102–121.

Pardeck, J., Umfress, T., Collier, K., and Murphy, J. W. 1987. The use and perception of computers by professional social workers. *Family Therapy* 14 (1): 1–8.

Patton, M. J. 1978. *Utilization-Focused Evaluation*. Beverly Hills, CA: Sage.

Patterson, D. and Yaffe, J. 1993. Using computer-assisted instruction to teach Axis II of the *DSM–III–R* to social work students. *Research on Social Work Practice* 3:343–357.

Peak, T., Toseland, R. W., and Banks, S. M. 1995. The impact of a spouse-caregiver support group on care recipient health care costs. *Journal of Aging and Mental Health* 7 (3): 427–449.

Pear, R. 1999, June 8. N.I.H. plan for journal on the web draws fire. *The New York Times*, Science Times.

Peile, C. 1988. Research paradigms in social work: From stalemate to creative synthesis. *Social Service Review* 62 (1): 2–19.

Pelz, D. C. 1978. Some expanded perspectives on the use of social science in public policy. In M. Yinger and S. J. Cutler, eds., *Major Social Issues: A Multidisciplinary View*, 346–357. New York: Free Press.

Penka, C. E. and Kirk, S. A. 1991. Practitioner involvement in clinical evaluation. *Social Work* 36:513–518.

Penn, D. L. and Mueser, K. T. 1996. Research update on the psychosocial treatment of schizophrenia. *American Journal of Psychiatry* 153:607–617.

Pepper, S. 1970. *World Hypotheses*. Berkeley: University of California Press.

Perlman, H. 1957. *Social Casework: A Problem-Solving Process*. Chicago: University of Chicago Press.

——. 1972. Once more, with feeling. In E. J. Mullen and J. Dumpson, eds., *Evaluation of Social Intervention*, 191–209. London: Jossey-Bass.

Peuto, B. L. 1997, March 31. Mainframe history provides lessons. (Industry Trend or Event). *Microprocessor Report* 11 (4): 19–23.

Phillips, D. C. 1987. *Philosophy, Science and Social Inquiry*. New York: Pergamon.

——. 1992. *The Social Scientist's Bestiary: A Guide to Fabled Threats to, and Defenses of, Naturalistic Social Science*. New York: Pergamon.

Polansky, N. A. 1960. *Social Work Research*. Chicago: University of Chicago Press.

——. 1971. Research in social work. In R. Morris, ed., *Encyclopedia of Social Work 1971*, 16th ed., 1098–1106. New York: National Association of Social Workers.

——. 1977. Research in social work: Social treatment. In J. B. Turner, ed., *Encyclopedia of Social Work 1977*, 17th ed., 1206–1213. Washington, DC: National Association of Social Workers.

Popper, K. 1959. *The Logic of Scientific Discovery*. London: Hutchinson.

Powers, E. and Witmer, H. 1951. *An Experiment in the Prevention of Delinquency— The Cambridge-Somerville Youth Study*. New York: Columbia University Press.

Press, L. 1993. Before the Altair: The history of personal computing. *Communications of the ACM* 36 (9) (Sept. 27).

Proctor, E. and Rosen, A. 1999. E-mail correspondence, 2 December.

——. In press. The structure and function of social work practice guidelines. In E. Proctor and A. Rosen, eds., *Developing Practice Guidelines for Social Work Intervention*. New York: Columbia University Press.

Pumpian-Mindlin, E. 1952. The position of psychoanalysis in relation to the biological and social sciences. In E. Pumpian-Mindlin, ed., *Psychoanalysis as Science*. New York: Basic.

Random House Unabridged Dictionary, 2nd ed. New York: Random House, 1993.

Raushi, T. M. 1994. *A Task-Centered Model for Group Work with Single Mothers in the College Setting*. Ph.D. diss., State University of New York at Albany.

Reamer, F. G. 1992. *Philosophical Foundations of Social Work*. New York: Columbia University Press.

——. 1990. The nature of expertise in social welfare. In L. Videka-Sherman and W. J. Reid, eds., *Advances in Clinical Social Work Research*, 88–91. Washington, DC: National Association of Social Workers.

Rehr, H. 1992. Research utilization and applications of findings. *Research in Social Work Practice* 2 (3): 358.

Reid, W. J. 1974. Developments in the use of organized data. *Social Work* 19 (5): 585–593.

——. 1975. A test of a task-centered approach. *Social Work* 20:3–9.

——. 1978a. The social agency as a research machine. *Journal of Social Service Research* 2 (1): 11–23.

——. 1978b. *The Task-Centered System*. New York: Columbia University Press.

——. 1979. The model development dissertation. *Journal of Social Service Research* 3:215–225.

——. 1983. Research developments. In S. Briar, ed., *1983 Supplement to the Encyclopedia of Social Work*, 17th ed., 128–134. Washington, DC: National Association of Social Workers.

——. 1985a. *Family Problem Solving*. New York: Columbia University Press.

——. 1985b. Task-centered treatment. In F. J. Turner, ed., *Social Work Treatment: Interlocking Theoretical Approaches*, 3rd ed., 267–295. New York: Free Press, 1979.

——. 1987. Evaluating an intervention in developmental research. *Journal of Social Service Research* 11:17–39.

Reid, W. J. 1994a. The empirical practice movement. *Social Service Review* 68:165–184.

——. 1994b. Field testing and data gathering on innovative practice interventions in early development. In J. Rothman and E. J. Thomas, eds., *Intervention Research*, 245–264. New York: Haworth.

——. 1994c. Reframing the epistemological debate. In E. Sherman and W. J. Reid, eds., *Qualitative Research in Social Work*, 464–481. New York: Columbia University Press.

——. 1997a. Evaluating the dodo's verdict: Do all interventions have equivalent outcomes? *Social Work Research* 21:5–18.

——. 1997b. Long-term trends in clinical social work. *Social Service Review* 71 (2): 200–213.

——. 1997c. Research on task-centered practice. *Social Work Research* 21 (3): 132–137.

Reid, W. J. and Bailey-Dempsey, C. 1994. Content analysis in design and development. *Research on Social Work Practice* 4:101–114.

——. 1995. The effects of monetary incentives on school performance. *Families in Society* 76:331–340.

Reid, W. J. and Davis, I. 1987. Qualitative methods in single case research. In N. Gottlieb, ed., *Proceedings of Conference on Practitioners as Evaluators of Direct Practice*, 56–74. Seattle, WA: Center for Social Welfare Research, School of Social Work, University of Washington.

Reid, W. J. and Epstein, L., eds. 1972. *Task-Centered Casework*. New York: Columbia University Press.

——. 1977. *Task-Centered Practice*. New York: Columbia University Press.

Reid, W. J., Epstein, L., Brown, L. B., Tolson, E. R., and Rooney, R. H. 1980. Task-centered school social work. *Social Work in Education* 2:7–24.

Reid, W. J. and Fortune, A. E. 1992. Research utilization in direct social work practice. In A. Grasso and I. Epstein, eds., *Research Utilization in the Social Services*, 97–115. New York: Haworth.

——. 2000. Harbingers, origins, and approximations of evidence-based practice in current social work knowledge. Paper presented at Toward the development of evidence-based practice for social work intervention: A working conference, George Warren Brown School of Social Work, Washington University, St. Louis, MO.

Reid, W. J. and Hanrahan, P. 1982. Recent evaluations of social work: Grounds for optimism. *Social Work* 27:328–340.

Reid, W. J. and Shyne, A. 1969. *Brief and Extended Casework*. New York: Columbia University Press.

Reid, W. J. and Smith, A. 1989. *Research in Social Work*, 2nd ed. New York: Columbia University Press.

Reid, W. J. and Zettergren, P. 1999. Empirical practice in evaluation in social work practice. In I. Shaw and J. Lishman, eds., *Evaluation in Social Work Practice*, 41–62. Thousand Oaks, CA: Sage.

Rich, M. E. 1926. Editorial. *The Family* 7:247.

Rich, R. F. 1977. Uses of social science information by federal bureaucrats: Knowledge for action versus knowledge for understanding. In C. H. Weiss, ed., *Using Social Research in Public Policy Making*. Lexington, MA: Lexington Books.

Richey, C. A., Blythe, B. J., and Berlin, S. B. 1987. Do social workers evaluate their practice? *Social Work Research and Abstracts* 23:14–20.

Richey, C. A. and Roffman, R. A. 1999. On the sidelines of guidelines: Further thoughts on the fit between clinical guidelines and social work practice. *Research on Social Work Practice* 9:311–321.

Richmond, M. 1917. *Social Diagnosis*. New York: Russell Sage Foundation.

Rimer, E. 1986. Implementing computer technology in human service agencies: The experience of two California counties. *New England Journal of Human Services* 6 (3): 25–29.

Ripple, L., Alexander, E., and Polemis, B. W. 1964. *Motivation, Capacity and Opportunity*. Chicago: University of Chicago Press.

Roberts, A. O. H. and Larsen, J. K. 1971. *Effective Use of Mental Health Research Information*. Palo Alto, CA: American Institute for Research.

Robinson, E. A. R., Bronson, D. E., and Blythe, B. J. 1988. An analysis of the implementation of single-case evaluation by practitioners. *Social Service Review* 62:285–301.

Robinson, L. A., Berman, J. S., and Neimeyer, R. A. 1990. Psychotherapy for the treatment of depression: A comprehensive review of controlled outcome research. *Psychological Bulletin* 100:30–49.

Robinson, V. P. 1921. Analysis of processes in the records of family caseworking agencies. *The Family* 2 (5): 101–106.

Rock, B. and Congress, E. 1999. The new confidentiality for the 21st century in a managed care environment. *Social Work* 44 (3): 253–262.

Rodwell, M. K. 1998. *Social Work Constructivist Research*. New York: Garland.

Rooney, R. H. 1978. *Separation Through Foster Care: Toward a Problem-Oriented Practice Model Based on Task-Centered Casework*. Ph.D. diss., University of Chicago.

——. 1981. A task-centered reunification model for foster care. In A. N. Malluccio and P. A. Sinanoglu, eds., *The Challenge of Partnership: Working with Parents of Children in Foster Care*. New York: Child Welfare League of America.

——. 1988. Measuring task-centered training effects on practice: Results of an audiotape study in a pubic agency. *Journal of Continuing Social Work Education* 42:2–7.

Root, L. 1996. Computer conferencing in a decentralized program: An occupational social work example. *Administration in Social Work* 20:31–45.

Rorty, R. 1979. *Philosophy and the Mirror of Nature*. Princeton: Princeton University Press.

Rosen, A. 1994. Knowledge use in direct practice. *Social Service Review* 68:561–577.

Rosen, A. and Mutschler, E. 1982. Social work students' and practitioners' orientation to research. *Journal of Education for Social Work* 18:62–68.

Rosen, A., Proctor, E., and Staudt, M. 1999. Social work research and the quest for effective practice. *Social Work Research* 23:4–14.

Rosenblatt, A. 1968. The practitioner's use and evaluation of research. *Social Work* 13:53–59.

Rosenthal, R. 1984. *Meta-Analytic Procedures for Social Research*. Beverly Hills, CA: Sage.

Rossi, P. H. 1977. Research in social work: Social policy. In J. B. Turner, ed., *Encyclo-*

pedia of Social Work 1977, 17th ed., 1202–1206. Washington, DC: National Association of Social Workers.

Rothchild, A. W. and Bedger, J. E. 1974. A childata system can work. *Child Welfare* 53:1.

Rothman, J. 1974. *Planning and Organizing for Social Change: Action Principles from Social Research*. New York: Columbia University Press.

——. 1980. *Research and Development in the Human Services*. Englewood Cliffs, NJ: Prentice-Hall.

——. 1989. Intervention research: Application to runaway and homeless youths. *Social Work Research and Abstracts* 25 (1): 13–18.

——. 1992. *Guidelines for Case Management: Putting Research to Professional Use*. Itasca, IL: Peacock.

Rothman, J. et al. 1983. *Marketing Human Service Innovations*. Beverly Hills, CA: Sage.

Rothman, J. and Teresa, J. G. 1978. *Fostering Participation and Promoting Innovation*. Itasca, IL: Peacock.

Rothman, J. and Thomas, E. J., eds. 1994. *Intervention Research: Design and Development for Human Service*. New York: Haworth.

Rothman, J. and Tumblin, A. 1994. Pilot testing and early development of a model of case management intervention. In J. Rothman and E. J. Thomas, eds., *Intervention Research: Design and Development for Human Service*, 215–243. New York: Haworth.

Rubin, A. 1985. Practice effectiveness: More grounds for optimism. *Social Work* 30:469–476.

——. 1999. Presidential editorial: Do National Association of Social Workers leaders value research? A summit follow-up. *Research on Social Work Practice* 9 (3): 277–282.

Rubin, A. and Knox, K. S. 1996. Data analysis problems in single-case evaluation: Issues for research on social work practice. *Research on Social Work Practice* 6 (1): 40–65.

Rubin, A. and Rosenblatt, A., eds. 1979. *Sourcebook on Research Utilization*. New York: Council on Social Work Education.

Rubin, E. R. 1976. The implementation of an effective computer system. *Social Casework* 57 (7): 438–445.

Runyan, L. 1991, March 15. 40 years on the frontier. *Datamation* 37 (6): 34–47.

Schilling, R. F., Schinke, S. P., and Gilchrist, L. D. 1985. Utilization of social work research: Reaching the practitioner. *Social Work* 30:527–529.

Schoech, D. 1995. Information systems. In R. L. Edwards, ed., *Encyclopedia of Social Work 1995*, 19th ed., 1470–1479. Washington, DC: National Association of Social Workers.

Schoech, D. and Arangio, T. 1979. Computers in the human services. *Social Work* 24 (2): 96–102.

Schoech, D., Jennings, H., Schkade, L. L., and Hooper-Russell, C. 1985. Expert systems: Artificial intelligence for professional decisions. *Computers in Human Services* 1 (1): 81–115.

Schoech, D. and Schkade, L. L. 1980. Computers helping caseworkers: decision support systems. *Child Welfare* 59 (9): 566–575.

Schoech, D., Schkade, L., and Myers, R. 1982. Strategies for information system development. *Administration in Social Work* 5 (3/4): 25–26.

Schopler, J. H., Abell, M. D., and Galinsky, M. J. 1998. Technology-based groups: A review and conceptual framework for practice. *Social Work* 43 (3): 254–267.

Schopler, J. H., Galinsky, M. J., and Abell, M. 1997. Creating community through telephone and computer groups: Theoretical and practice perspectives. *Social Work with Groups* 20 (4): 19–34.

Schuerman, J. R. 1982. The obsolete scientific imperative in social work research. *Social Service Review* 56:144–148.

——. 1987. Expert consulting systems in social welfare. *Social Work Research and Abstracts* 23 (3): 14–18.

Schuerman, J. R., Mullen, E., Stagner, M., and Johnson, P. 1988. First generation expert systems in social welfare. *Computers in Human Services* 4 (1/2): 111–122.

Schwab, A. J., Jr., Bruce, M. E., and McRoy, R. G. 1985. A statistical model of child placement decisions. *Social Work Research and Abstracts* 21 (2): 28–34.

——. 1986. Using computer technology in child placement decisions. *Social Casework: The Journal of Contemporary Social Work* 67 (6): 359–368.

Schwartz, E. E. and Sample, W. C. 1967. First findings from Midway. *Social Service Review* 41:113–151.

——. 1972. *The Midway Office.* New York: National Association of Social Workers.

Seligman, M. 1995. The effectiveness of psychotherapy: The consumer reports study. *American Psychologist* 50:965–974.

Semke, J. I. and Nurius, P. S. 1991. Information structure, information technology, and the human services organizational environment. *Social Work* 36 (4): 353–358.

Shaw, I. and Shaw, A. 1997. Game plans, buzzes, and sheer luck: Doing well in social work. *Social Work Research* 21 (2): 69–79.

Sheffield, A. E. 1937. *Social Insight in Case Situations.* New York: D. Appleton-Century.

Sheldon, B. 1986. Social work effectiveness experiments: Review and implications. *British Journal of Social Work* 16:233–242.

Shyne, A. W. 1965. Social work research. In H. L. Lurie, ed., *Encyclopedia of Social Work 1965*, 15th ed., 763–772. New York: National Association of Social Workers.

Siegel, D. H. 1983. Can research and practice be integrated in social work education? *Journal of Education for Social Work* 19 (3): 12–19.

——. 1985. Effective teaching of empirically based practice. *Social Work Research and Abstracts* 21:40–48.

Siegel, H. 1987. *Relativism Refuted.* Dordrecht, The Netherlands: Reidel.

Slonim-Nevo, V. and Anson, Y. 1998. Evaluating practice: Does it improve treatment outcome? *Social Work Research* 22 (2): 66–75.

Smith, C., Lizotte, A. J., Thornberry, T. P., and Krohn, M. D. 1995. Resilient youth: Identifying factors that prevent high-risk youth from engaging in delinquency and drug use. *Current Perspectives on Aging and the Life Cycle* 4:217–247.

Smith, M. L., Glass, G. V., and Miller, T. I. 1980. *The Benefits of Psychotherapy*. Baltimore: Johns Hopkins University Press.

Spencer, D. D. 1999. *The Timetable of Computers: A Chronology of the Most Important People and Events in the History of Computers*. Ormond Beach, FL: Camelot.

Spitzer, R. L. and Williams, J. B. W. 1983. Classification in psychiatry. In H. I. Kaplan and B. J. Sadock, eds., *Comprehensive Textbook of Psychiatry*, 591–613. Baltimore: Williams and Wilkins.

Spitzer, R. L. and Wilson, P. T. 1968. A guide to the American Psychiatric Association's new diagnostic nomenclature. *American Journal of Psychiatry* 124:1616–1629.

Stagner, M. 1994. Understanding and representing human service knowledge: The process of developing expert systems. *Journal of Social Service Research* 19 (1/2): 115–137.

Stalker, C. A., Levene, J. E., and Coady, N. 1999. Solution-focused brief therapy—one model fits all? *Families in Society* 80 (5): 468–478.

Staller, K. and Kirk, S. A. 1998. Knowledge utilization in social work and legal practice. *Journal of Sociology and Social Welfare* 25:93–115.

Starr, P. 1982. *The Social Transformation of American Medicine*. New York: Basic.

Stein, H. D. and Cloward, R. A. 1958. *Social Perspectives on Behavior*. Glencoe, IL: Free Press.

Stein, T. J. 1983. *Decision Making at Child Welfare Intake: A Handbook for Social Workers*. New York: Child Welfare League of America.

——. 1990. Issues in the development of expert systems to enhance decision making in child welfare. In L. Videka-Sherman and W. J. Reid, eds., *Advances in Clinical Social Work Research*, 84–87. Washington, DC: National Association of Social Workers.

Stein, T. J., Gambrill, E. D., and Wiltse, K. T. 1978. *Children in Foster Homes: Achieving Continuity in Care*. New York: Praeger.

Stein, T. J. and Rzepnicki, T. L. 1984. *Decision Making in Child Welfare Services*. Boston: Kluwer-Nijhoff.

Strassman, P. A. 1997. 40 years of IT history. *Datamation Magazine* 43 (10): 80–86.

Strauss, A. and Corbin, J. 1990. *Basics of Qualitative Research: Grounded Theory Procedures and Techniques*. Newbury Park, CA: Sage.

Stuart, R. B. 1968. Selecting a behavioral alternative through practice. *Group Psychotherapy* 21 (4): 219–222.

——. 1971. Research in social work: Social casework and social group work. In R.

Morris, ed., *Encyclopedia of Social Work 1971*, 16th ed., 1106–1122. New York: National Association of Social Workers.

Sullivan, R. J. 1980. Human issues in computerized social services. *Child Welfare* 59 (7): 401–406.

Sundel, M., Butterfield, W., and Geis, G. 1969. Modification of verbal behavior in chronic schizophrenics. *Michigan Mental Health Research Bulletin* 3 (2): 37–40.

Taber, M. A. and DiBello, L. V. 1990. The personal computer and the small social service agency. *Computers in Human Services* 6 (1/2/3): 181–197.

Taft, J. 1937. The relation of function to process in social casework. *Journal of Social Process* 1:1–18.

Task Force on Social Work Research. 1991. *Building Social Work Knowledge for Effective Services and Policies: A Plan for Research Development*. Austin, TX: Capital Printing Co.

Teicher, M. 1967. Social casework—science or art? *Child Welfare* 46:394–395.

Theis, S. V. 1924. *How Foster Children Turn Out*. New York: State Charities Aid Society.

Thomas, E. J. 1967a. Selecting knowledge from behavioral science. In E. J. Thomas, ed., *Behavioral Science for Social Workers*. New York: Free Press.

——, ed. 1967b. *The Socio-Behavioral Approach and Applications to Social Work*. New York: Council on Social Work Education.

——. 1978a. Generating innovation in social work: The paradigm of developmental research. *Journal of Social Service Research* 2:95–115.

——. 1978b. Mousetraps, developmental research, and social work education. *Social Service Review* 52:468–483.

——. 1978c. Research and service in single-case experimentation: Conflicts and choices. *Social Work Research and Abstracts* 14:20.

——. 1984. *Designing Interventions for the Helping Professions*. Beverly Hills, CA: Sage.

——. 1985. The validity of design and development and related concepts in developmental research. *Social Work Research and Abstracts* 21:50–55.

——. 1994. Appendix B: The unilateral treatment program for alcohol abuse background, selected procedures, and case applications. In J. Rothman and E. J. Thomas, eds., *Intervention Research: Design and Development for Human Service*, 427–447. New York: Haworth.

Thomas, E. J., Bastien, J., Stuebe, D., Bronson, D., and Yaffe, J. 1982. *The Critical Incident Technique: A Method of Assessing Procedural Adequacy*. Unpublished manuscript. University of Michigan School of Social Work.

Thomas, E. J. and Rothman, J., eds. 1994. An integrative perspective on intervention research. In J. Rothman and E. J. Thomas, eds., *Intervention Research: Design and Development for Human Service*, 3–23. New York: Haworth.

Thomas, E. J., Santa, C., Bronson, D., and Oyserman, D. 1987. Unilateral family therapy with the spouses of alcoholics. *Journal of Social Service Research* 10:145–162.

Thomas, E. J., Yoshioka, M. R., Ager, R., and Adams, K. B. 1990. Experimental outcomes of spouse intervention to reach the uncooperative alcohol abuser: preliminary report. In *Proceedings of the International Society for Bio-Medical Research and the Research Society on Alcoholism*, June 17–22, 1990. Toronto, Canada.

Tomm, K. 1982. Towards a cybernetic systems approach to family therapy. In F. W. Kaslow, ed., *The International Book of Family Therapy*, 70–90. New York: Brunner/Mazel.

Thyer, B. A. 2000. *Developing Discipline-Specific Knowledge for Social Work: Is it Possible?* Unpublished manuscript.

Thyer, B. A. and Thyer, K. B. 1992. Single-system research designs in social work practice: A bibliography from 1965 to 1990. *Research on Social Work Practice* 2 (1): 99–116.

Thyer, B. A. and Wodarski, J. S., eds. 1998. *Handbook of Empirical Social Work Practice: Volume I: Mental Disorders*. New York: Wiley.

Tolson, E. R. 1990. Synthesis: Why don't practitioners use single-subject designs? In L. Videka-Sherman and W. J. Reid, eds., *Advances in Clinical Social Work Research*, 58–64. Silver Springs, MD: National Association of Social Workers.

Toseland, R. W. and Reid, W. J. 1985. Using rapid assessment instruments in a family service agency. *Social Casework* 66: 547–555.

Tripodi, T. 1974. *Uses and Abuses of Social Research in Social Work*. New York: Columbia University Press.

Tripodi, T., Fellin, P., and Meyer, H. J. 1969a. *The Assessment of Social Research: Guidelines for Use of Research in Social Work and Social Science*. Itasca, IL: Peacock.

——. 1969b. *Exemplars of Social Research*. Itasca, IL: Peacock.

——. 1983. *The Assessment of Social Research: Guidelines for Use of Research in Social Work and Social Science*, 2nd ed. Itasca, IL: Peacock.

Tucker, D. J. 1996. Eclecticism is not a free good: Barriers to knowledge development in social work. *Social Service Review* 70:400–434.

Tyson, K. B. 1992. A new approach to relevant scientific research for practitioners: The heuristic paradigm. *Social Work* 37:541–556.

Vasey, I. T. 1968. Developing a data storage and retrieval system. *Social Casework* 49 (7): 414–417.

Velasquez, J. 1992. GAIN: A locally based computer system which successfully supports line staff. *Administration in Social Work* 16 (1): 41–54.

Velasquez, J. S., Kuechler, D. F., and White, M. S. 1986. Use of formative evaluation in a human service department. *Administration in Social Work* 10:67–77.

Videka-Sherman, L. 1988. Meta-analysis of research on social work practice in mental health. *Social Work* 33:325–338.

——. In press. Accounting for variability in client, population and setting characteristics: Moderators of intervention effectiveness. In E. Proctor and A. Rosen, eds., *Developing Practice Guidelines for Social Work Intervention*. New York: Columbia University Press.

Videka-Sherman, L. and Reid, W. J., eds. 1990. *Advances in Clinical Social Work Research*. Washington, DC: National Association of Social Workers.

Visser, A. 1997. Case-based learning, towards a computer tool for learning with cases. *New Technology in the Human Services* 10 (4): 11–14.

Vondracek, F. W. 1974. Feasibility of an automated intake procedure for human service workers. *Social Service Review* 48 (2): 271–278.

Wakefield, J. C. 1990. Expert systems, Socrates, and the philosophy of mind. In L. Videka-Sherman and W. J. Reid, eds., *Advances in Clinical Social Work Research*, 92–100. Washington, DC: National Association of Social Workers.

Wakefield, J. C. and Kirk, S. A. 1996. Unscientific thinking about scientific practice: Evaluating the scientist-practitioner model. *Social Work Research* 20 (2): 83–95.

Wallace, D. 1967. The Chemung County evaluation of casework service to dependent multi-problem families: Another problem outcome. *Social Service Review* 41:379–389.

Warner, A. G. 1894. *American Charities: A Study in Philanthropy and Economics*. New York: Thomas Y. Crowell.

Watson, F. D. 1922. *The Charity Organization Movement in the United States: A Study in American Philanthropy*. New York: Macmillan.

Watzlawick, P. 1967. A review of the double-bind theory. *Family Process* 2:132–153.

Watzlawick, P., Beavin, J., and Jackson, D. 1967. *Pragmatics of Human Communication*. New York: Norton.

Weed, P. and Greenwald, S. R. 1973. The mystics of statistics. *Social Work* 18:113–115.

Weinbach, R. W. 1984. Implementing change: Insights and strategies for the supervisor. *Social Work* 29:282–286.

Weinbach, R. and Rubin, A., eds. 1980. *Teaching Social Work Research: Alternative Programs and Strategies*. New York: Council on Social Work Education.

Weinberger, J. 1995. Common factors aren't so common: The common factors dilemma. *Clinical Psychology* 2:45–69.

Weiss, B. and Weisz, J. R. 1995. Relative effectiveness of behavioral versus nonbehavioral child psychotherapy. *Journal of Conslting and Clinical Psychology* 63 (2): 317–320.

Weiss, C. 1972. *Evaluating Action Programs: Readings in Social Action and Education*. Boston: Allyn and Bacon.

Weiss, C. H. and Bucavalas, M. J. 1980. Truth tests and utility tests: Decision-makers' frames of reference for social science research. *American Sociological Review* 45:302–313.

Wilensky, H. 1964. The professionalization of everyone? *American Journal of Sociology* 70:137–158.

Williams, J. B. W. and Lanigan, J. 1999. Practice guidelines in social wok: A reply, or our glass is half full. *Research on Social Work Practice* 9 (3): 338–342.

Wilson, M. 1993. *DSM–III*: The transformation of American psychiatry: A history. *American Journal of Psychiatry* 150:399–410.

Witkin, S. L. 1989. Towards a scientific social work. *Journal of Social Service Research* 12:83–98.

———. 1991a. Empirical clinical practice: A critical analysis. *Social Work* 36:158–165.

———. 1991b. The implications of social constructionist perspective. *Journal of Teaching in Social Work* 4:37–48.

———. 1996. If empirical practice is the answer, then what is the question? *Social Work Research* 20 (2): 69–75.

———. 1999. Constructing our future. *Social Work* 44 (1): 5–8.

Witkin, S. L., Edleson, J., and Lindsey, D. 1980. Social workers and statistics: Preparation, attitudes, and knowledge. *Journal of Social Service Research* 3:313–322.

Witkin, S. L. and Gottschalk, S. 1988. Alternative criteria for theory evaluation. *Social Service Review* 62 (2): 211–224.

Wodarski, J. S. 1987. Development of management information systems for human services: A practical guide. *Computers in Human Services* 3 (1/2): 37–49.

Wolmark, A. and Sweezy, M. 1998. Kohut's self psychology. In R. A. Dorfman, ed., *Paradigms of Clinical Social Work*, 45–69. New York: Brunner/Mazel.

Wood, K. M. 1978. Casework effectiveness: A new look at the research evidence. *Social Work* 23:437–458.

———. 1990. The family of the juvenile delinquent. *Juvenile and Family Court Journal* 11:19–37.

Zaslaw, M. J. 1988. Sex differences in children's response to parental divorce: 1. Research methodology and postdivorce family forms. *American Journal of Orthopsychiatry* 59:355–378.

———. 1989. Sex differences in children's response to parental divorce: 2. Samples, variables, ages, and sources. *American Journal of Orthopsychiatry* 59:118–140.

Zeira, A. and Rosen, A. 2000. Unraveling "tactic knowledge": What social workers do and why they do it. *Social Service Review* 74:103–123.

Zilborg, G. 1941. *A History of Medical Psychology*. New York: Norton.

Zimbalist, S. E. 1977. *Historic Themes and Landmarks in Social Welfare Research*. New York: Harper and Row.

Zuckerman, H. and Merton, R. K. 1971. Patterns of evaluation in sciences: Institutionalization, structure and functions of the referee system. *Minerva* 1:66–200.